T

General Editors
PATRICIA WAUGH AND LYNNE PEARCE

Titles in the INTERROGATING TEXTS series

Already published:

Practising Postmodernism/Reading Modernism
Patricia Waugh

Reading Dialogics
Lynne Pearce

Critical Desire
Psychoanalysis and the Literary Subject
Linda R. Williams

Forthcoming:

Essentially Plural
Feminist Literary Theory
Marion Wynne-Davies

Theorizing the Fantastic
Lucie Armitt

Interrogating History/Reading Ourselves
New Historicism, Cultural Materialism, and Marxism in the Contemporary Literary Debate
Jeremy Hawthorn

CRITICAL DESIRE

Psychoanalysis *and the* Literary Subject

Linda Ruth Williams
Lecturer, University of Southampton

Edward Arnold
A member of the Hodder Headline Group
LONDON NEW YORK SYDNEY AUCKLAND

For Irene, George and Derrick Williams

First published in Great Britain 1995 by
Edward Arnold, a division of Hodder Headline PLC,
338 Euston Road, London NW1 3BH
175 Fifth Avenue, New York, NY 10010

Distributed exclusively in the USA by
St. Martin's Press, Inc.
175 Fifth Avenue, New York, NY 10010

British Library Cataloguing in Publication Data
A catalogue record for this book is available from the British Library

Library of Congress Cataloging-in-Publication Data
Williams, Linda Ruth.
Critical desire: psychoanalysis and the literary subject/
Linda Ruth Williams.
p. cm. — (Interrogating texts)
Includes bibliographical references and index.
ISBN 0–340–64557–1. — ISBN 0–340–56816–X (pbk.)
1. Psychoanalysis and literature. I. Title. II. Series.
PN56.P92W55 1995
801'.92—dc20 94–48336
CIP

ISBN 0 340 64557 1 (hb)
ISBN 0 340 56816 X (pb)

1 2 3 4 5 95 96 97 98 99

Typeset in 10 on 12pt Palatino by
Phoenix Photosetting, Chatham, Kent
Printed and bound in Great Britain by
J W Arrowsmith Ltd, Bristol

Contents

General Editors' Preface

Interrogating Texts is a series which aims to take literary theory – its key proponents, debates, and textual practices – towards the next century.

As editors we believe that despite the much vaunted 'retreat from theory', there is so far little material evidence of this supposed backlash. Publishers' catalogues reveal 'theory' (be it literary, cultural, philosophical or psychoanalytic) to be an expanding rather than a contracting market, and courses in literary theory and textual practice have now been established in most institutions of Higher Education throughout Europe and North America.

Despite significant improvements to high school syllabuses in recent years, however, most students still arrive at University or College ill-prepared for the 'revolution' that has shaken English studies in the past twenty years. Amid the welter of increasingly sophisticated and specialized critical works that now fill our libraries and bookshops, there is a pressing need for volumes like those represented by this series: volumes that will summarize, contextualize and *interrogate* the key debates informing contemporary literary theory and, most importantly, assess and demonstrate the *effectiveness* of the different approaches in the reading of literary texts.

It is, indeed, in its 'conceptual' approach to theory, and its 'interrogation' of theory *through* textual practice, that the series claims to be most strikingly new and distinctive. Instead of presenting literary theory as a series of 'approaches' (eg., Structuralism, Marxism, Feminism) that can be mechanistically 'applied' to any text, each volume will begin by examining the epistemological and conceptual frameworks of the theoretical discourse in question and examine the way in which its philosophical and political premises compare and contrast with those of other contemporary discourses. (The volumes on *Postmodernism* and *Dialogics* both consider their epistemological relation to the other, for example.) Each volume, too, will provide a historical overview of the key proponents, texts, and debates represented by the theory, as well as an evaluative survey of the different ways in which the theory has been appropriated and deployed by literary critics. Alongside this informative and evaluative contextualization of the theory, each volume will perform readings of a selection of literary texts. The aim of these readings, as indicated earlier, is not simplistically to demonstrate the way in which the theory in question can be 'applied' to a text, but to question the suitability of certain aspects of the theory *vis-à-vis* certain texts, and ultimately to use the texts to *interrogate the theory itself*: to reveal its own inadequacies, limitations and blindspots.

Two of the most suggestive theoretical keywords of the 1980s were *dialogue* and *difference*. The Interrogating Texts series aims to (re)activate both terms in its attempt to map the great shifts and developments (the 'continental drift'?!) of literary theory over the past twenty years and into the twenty-first century: the differences both within and between the various theoretical discourses, and the dialogues that inhere and connect them.

Eschewing the mechanical association between theory and practice, it should also be pointed out that the individual volumes belonging to the series do not conform to any organizational template. Each author has been allowed to negotiate the relationship between theory and text as he or she thinks best, and in recognition of the fact that some of our theoretical categories will require a very different presentation to others.

Although both the substance and the critical evolution of the theoretical discourses represented by this series are often extremely complex, we hope that the perspectives and interrogations offered by our authors will make them readily accessible to a new generation of readers. The 'beginnings' of literary theory as a revolutionary threat and disruption to the Academy is fast receding into history, but its challenge – what it offers each of us in our relentless interrogation of literary texts – lives on.

Lynne Pearce
Patricia Waugh

Preface

At the beginning of section 5 of the 'Wolf Man' case, Freud presents us with an image of estranged worlds: 'The whale and the polar bear, it has been said, cannot wage war on each other, for since each is confined to his own element they cannot meet.'[1] This model of different languages might stand for how the difference between literature and psychoanalysis has been perceived. It is the purpose of this book not only to make the whale and the polar bear meet, but to show that their languages can be (indeed, are always being) intriguingly translated into each other's. '[T]heoretical controversy is unfruitful,' continues Freud; 'it seems to me to be far more useful to combat dissentient interpretations by testing them upon particular cases and problems.' With such practical advice in mind, this book will offer a meeting of psychoanalysis with the 'particular cases and problems' of some literary texts and films, in order to explore what each can say about and to the other, as well as what they *cannot*.

I have tried to juxtapose a range of materials from different genres and periods with as substantial a selection of psychoanalytic theories as was practical here, but this has clearly meant imposing certain limits, which have, I hope, allowed me to do what I *am* doing better. This book is Freudian in its theoretical emphases, concentrating on debates emanating from the work of Freud, Lacan, and their readers. I have not, for instance, looked at Jung or Jungian criticism at all, and I have only discussed psychoanalysis as a therapeutic practice when this has directly informed my reading of it as a body of thought. Chapter 1 does, however, establish how Freud's models of the unconscious, infant sexuality, repression and the analysand-as-text emerged through his abandonment of the seduction theory and his development of free-association

1 Sigmund Freud, 'From the History of an Infantile Neurosis (The 'Wolf Man')' (1918), in *Case Histories II*, trans. James Strachey, ed. Angela Richards, Pelican Freud Library vol. 9 (Harmondsworth: Penguin, 1981), p. 281.

techniques. I then take this through into an investigation of the case history as story (Freud's *Dora*) and the story as case history (Elizabeth Gaskell's *Cousin Phillis*), before moving on, in Chapter 2, to a direct discussion of some basic debates (circulating around the work of Edgar Allan Poe) in psychoanalytic literary criticism since Freud, mediated by Marie Bonaparte, Jacques Lacan, Jacques Derrida, Shoshana Felman and Barbara Johnson. In Chapter 3 I look more closely at psychoanalytic models of fantasy through the work of Angela Carter, and discuss how these might intersect with literary readings of fantasy as a genre. This involves a comparative discussion of Freud on fetishism, Jean Laplanche and J.-B. Pontalis on primal fantasy, and a short story by erotic writer Anaïs Nin, 'The Veiled Woman'. The chapter returns to Carter through a discussion of Klein, Winnicott and object relations theory, as well as the feminist-psychoanalytic 'return to the mother'.

Thus each of the chapters in this book focuses on a key psychoanalytic concept in some detail, by looking primarily at one or two texts. Chapter 4 encounters the disruption of linear time in the narrative of the subject, first through an analysis of Jean Laplanche's work on deferred action, 'afterwardsness' or *nachträglichkeit*. This is set up through a discussion of Freud's Wolf Man case and Mrs Oliphant's 1899 *Autobiography*, but in order to show the importance of this concept for our understanding of narrative, I also look at John Boorman's 1972 film *Deliverance*, in order to think about how the subject is constructed in cinematic time. In Chapter 5 I turn to a relatively neglected area, the psychoanalytic theory of the death drive, and its implications for readings of both gothic narratives and Christina Rossetti's lyric poetry. Finally, Chapter 6 encounters a text thick with resistances to psychoanalysis, Thomas Hardy's *Jude the Obscure*, which, I argue, becomes more intriguing precisely because of this resistance.

I hope that already implicit in the work I am doing here is a sense that the way forward lies *between* these disciplines. I want to offer one simple point about interdisciplinary work to frame some of the arguments of this text, as well as to suggest where things might be going beyond the scope of this book. In her Introduction to the recent collection of essays, *Psychoanalysis and Cinema*, E. Ann Kaplan writes,

> It is unfortunate that, historically, literary and film scholars have not shown more interest in each other's work: it is to be regretted that even very recent literature/psychoanalysis anthologies have not included essays on film: although this is understandable in terms of the disciplinary boundaries around which we continue to construct our scholarly activities (i.e. our journals, our conferences, our departments), it would seem that dialogue could benefit both groups.[2]

2 E. Ann Kaplan, 'Introduction', *Psychoanalysis and Cinema* (London and New York: Routledge, 1990), p. 1.

Rather than positing this greater dialogue between these three discourses as one way forward for psychoanalytic cultural theory at the *conclusion* of this book, I have tried to bring these together in the writing of it. Although my readings are predominantly literary, I have mixed these with some discussion of films as well as with specific consideration of film theory. This is not simply because intrinsic to the psychoanalytic discussion of the subject is an awareness of how it is constructed visually and through the image, but because film as well as literature is an important site upon which the cultural life of psychoanalysis is developing.

Acknowledgements

This book was written across a busy period which involved me in many changes. I would like to thank colleagues at the University of Liverpool for their support during the time I was working there, before I moved to my present job at Southampton University. Much of this was written during the period of study-leave I was granted by Liverpool, and I learned a lot from exploring these ideas with students on a number of Liverpool's courses, particularly the MA in Victorian Literature and the third year 'Visions of Excess' course. Simon Dentith, Nick Davis and Jocelyn Wogan-Browne showed particular interest in the project and offered useful advice, and I have benefited enormously from talking vampires and Rossetti with Anna Powell and Emma Francis (respectively) over the last few years. Thanks also to staff and graduate students at Cambridge and Lancaster Universities, where drafts of Chapters 3 and 5 were given at graduate forums, as well as at the 'Modernism' conference at Cambridge in 1993. Denise Riley's imaginative comments on Chapter 5 were also very helpful. An early version of Chapter 6 was presented at the Thomas Hardy Conference at the Universidade do Porto.

More personal thanks go to the two energetic souls who read the whole manuscript through in its final stages. Rob Harry Dean's witty and perceptive reading of it was invaluable. Discussions I have had with him over the last eight years on many aspects of this subject have been a constant inspiration – as has his friendship. Finally, I am extremely grateful to Mark Kermode for also working through the manuscript so meticulously, as well as for his love and support throughout. This book is for him too.

A version of the discussion of *Deliverance* now in Chapter 4 was first published in a different form in *Sight and Sound* in October 1994.

1

Freud, Hysteria and 'the Art of Interpretation'

We are born into stories. Everything we remember is remembered through narratives, verbal constructions, images which individual history imbues with the lustre of myth. Our scenarios of the past, our dreams and our texts, are also understood through the possibilities of interpreting them which living in a specific culture offers us. Telling, writing, interpreting these stories is the stuff of psychoanalysis as well as literature, which reads individual and unconscious narratives, verbal or imagistic vignettes laced with significance, in very specific ways. It is the work of this book to outline what some of these ways might be.[1] Particularly, I will be looking at the moments at which psychoanalysis and literature (as well as other cultural forms such as film) interconnect, with psychoanalysis itself seen as interdependent with the cultural forms it is often said simply to inform.

This first chapter will begin to suggest what it is about the psychoanalytic emphasis on the unconscious, sexuality and difference, the construction of the subject – the self understood not as an essentially determined entity which grows according to a God-given or biologically ordained pattern, but as something formed through a worldly, anxious process of

1 In this chapter I offer some basic accounts of key Freudian models and historical moves where appropriate. Since readers will probably be familiar with some of these terms, I have kept these descriptions brief. The work of this book is to show the literary life of these models, and I have preferred to dedicate the space I have to opening this up. Those wishing for a fuller account of basic terms are strongly urged to read the primary Freud texts which I refer to here. Freud's historical and expository essays, collected together in the easily available Pelican Freud vol. xv, *Historical and Expository Works on Psychoanalysis*, have much to recommend them as clear, accessible ways into the key developments of psychoanalytic history. I have used the Pelican Freud (PF) wherever possible.

development – which has made it so attractive to cultural analysis and the understanding of literary texts and films. But it will also show these apparently different fields as intimately bound together in their histories and practices. Psychoanalysis itself is exactly as old as cinema – both developed as distinct forms in the early to mid-1890s, although both also have roots in practices dating back much earlier in the nineteenth century. Psychoanalytic criticism pre-dates the establishment of English literature as a discrete subject,[2] and it has outlived a number of once-fashionable critical schools, being now one of the most dynamic areas of the subject. It is perhaps the oldest methodology still in use, continuing to forge one of the theoretical cutting-edges in literary and film studies. Just as the history of critical turns can be traced through a bare-bones account of psychoanalytic literary developments this century, so changes in psychoanalysis itself have had a profound impact on the way in which we read.

But there is a more intimate interconnection or interpenetration of psychoanalysis and literature, which is belied by their apparent separation into clinical and cultural practices, or social-scientific and artistic disciplines. Psychoanalytic literary criticism is often popularly understood as the crude mapping onto texts of pre-formed psychological models, and whilst this is still a prevalent procedure (which will be explored in detail in Chapter 2), it often assumes that the psychoanalytic models in use are themselves somehow pre- or essentially non-literary. So what ways of reading have formed around the 'text' of the speaking or dreaming subject of analysis? Moreoever, how does the subject come into being through language, images, and models, which are best understood through a form of analysis which is itself at root essentially 'literary'?

As well as looking at these questions through a discussion of the development of psychoanalysis in terms of different readings of hysteria, later in the chapter I will also set up a reading of Elizabeth Gaskell's short novel *Cousin Phillis* (itself of interest for the positions on hysteria it opens up), in order to avoid as far as possible separating psychoanalytic theory from literary practice, since the aim throughout this book will be to keep in play the interpenetration of psychoanalysis and other cultural forms. This chapter will explore the literary and linguistic acts at the heart and origin of psychoanalysis itself through an account of Freud's early work on hysteria, as well as showing directly the bearing these have on one short, accessible narrative.

2 As a University-taught discipline, English Literature came into being during the First World War and was only fully established with *Scrutiny* and Cambridge criticism in the 1920s and 1930s. For a full discussion of this see Terry Eagleton, 'The Rise of English', in *Critical Theory: An Introduction* (London and Basingstoke: Macmillan 1983), Francis Mulhern, *The Moment of Scrutiny* (London: New Left Books, 1979), and Chris Baldick, *The Social Mission of English Criticism* (Oxford: Clarendon, 1983).

I

'Illnesses which Speak':[3] Reading Hysteria

a case of hysteria is a caricature of a work of art
Sigmund Freud, *Totem and Taboo*[4]

Psychoanalysis begins with women, in the sense that it is predicated on the reading of hysteria. Charles Bernheimer's Introduction to *In Dora's Case: Freud – Hysteria – Feminism* (the 1985 collection of essays on the 'Dora' case) boldly begins, 'Freud invented psychoanalysis between 1895 and 1900 on the basis of his clinical experience with hysterical patients, nearly all of them women, and of the self-analysis he performed to cure his own hysterical symptoms'.[5] Hysteria is an illness with a long history, meticulously charted by Ilza Veith in her 1965 book, *Hysteria: The History of a Disease*,[6] dating back to Hippocrates and even the Ancient Egyptians. Commonly known as a woman's illness (the name comes from the Greek hysteros, 'womb'), what was in the late nineteenth century known by this name (it is a term, and a diagnosis, which has lost currency more recently) was characterized by a range of multiform symptoms which, because they were focused on the body itself, gave the appearance of a physical condition with a direct somatic cause which could nevertheless not be found. A patient may have uncontrollable fits, contortions, paralyses, pain, muscular rigidity, yet with no identifiable physical reason for these ostensibly somatic signs of unrest; the condition was something of a pathological movable feast, fixing on parts of the body apparently at random. It was this which gave rise to the notorious 'wandering womb theory' of hysteria: that the womb, an essentially malignant organ floating freely in the body, would move around and fix itself upon the hapless bodily location which then produced the particular symptom from which the hysteric would be suffering. Although this notion was largely discredited by the time of the 'Classical Age' of seventeenth-century medicine, Michel Foucault can nevertheless quote translator Jean Liébault expressing the view at that time:

3 Lacan, 'Intervention on Transference', in Lacan and the *École Freudienne, Feminine Sexuality* (London and Basingstoke: Macmillan, 1983), p. 63.

4 Sigmund Freud, 'Totem and Taboo', PF vol. xiii, p. 130. Since full references are given to these works under their titles in the Bibliography at the end of this volume, I shall confine references in footnotes to the title of the paper and the PF volume number.

5 See Charles Bernheimer and Claire Kahane, eds., *In Dora's Case: Freud – Hysteria – Feminism* (New York: Columbia University Press, 1985), p. 1.

6 Ilza Veith, *Hysteria: The History of a Disease* (Chicago: Chicago University Press, 1965). See also Elaine Showalter, *The Female Malady: Women, Madness and English Culture, 1830-1980* (London: Virago, 1987)

> And so the womb, though it be so strictly attached to the parts we have described that it may not change place, yet often changes position, and makes curious and so to speak petulant movements in the woman's body. These movements are various: to wit, ascending, descending, convulsive, vagrant, prolapsed. The womb rises to the liver, spleen, diaphragm, stomach, breast, heart, lung, gullet, and head.[7]

Although by the Nineteenth century medicine had rejected this notion, the sense in which the study of hysteria was about reading and interpreting the mysteries of women's bodies remained, since these were still seen as at the command of a wayward internal will (the nervous system). Inscribed in this quotation are also implicitly the hallmarks of a disorder which is itself 'various', curiously changing position, attacking body parts which suffer for no manifest local reason. Hysteria displays, gives out, puts on show a 'pantomime' of symptoms which were eventually to be read as external 'representations' of internal unrest.[8] Yet for all its ostentatious display of physical signs, its cause was as yet outwardly unplaceable. This was an illness in search of an interpreter.

Freud worked with French neurologist Jean-Martin Charcot from 1885-6 at the Salpêtrière, a large asylum for women in Paris, in which Charcot developed pioneering ways of treating hysteria based on the understanding that it must be viewed not as the sign of a malignant womb, or of a malingering woman, or of demon-possession or witchcraft (only some of the ways in which it had been interpreted in the past), but as a disorder of the nervous system, 'a far more stable clinical entity than its centuries-old reputation for protean indefiniteness and chameleon-like duplicity gave reason to think'.[9] In the 1890s Freud, working with Joseph Breuer, took this further. Building on its new status as an accredited disease, he identified its origin not in the nervous system but in psychic conflict rooted in an anxious, unresolved experience of infantile sexuality. As Juliet Mitchell evocatively puts it, 'Freud decided that these tales of sound and fury did signify something',[10] and psychoanalysis began to be developed to trace this 'signification'. Hysteria emerged, moreover, not as one single condition but as a range of possible neurotic

7 Quoted by Michel Foucault, *Madness and Civilization* (London: Tavistock, 1967), pp. 143–4.

8 In 'Some General Remarks on Hysterical Attacks' (1909 [1908]), Freud writes, 'When one carries out the psychoanalysis of a hysterical woman patient whose complaint is manifested in attacks, one soon becomes convinced that these attacks are nothing else but phantasies translated into the motor sphere, projected on to motility and portrayed as pantomime' (PF vol. x, p. 97). Allon White and Peter Stallybrass discuss the notion of hysteria as pantomime in 'Bourgeois Hysteria and the Carnivalesque', pp. 171–90 of *The Politics and Poetics of Transgression* (London: Methuen, 1986).

9 Lisa Appignanesi and John Forrester, *Freud's Women* (London: Weidenfeld & Nicolson, 1992), p. 63.

10 Juliet Mitchell, *Women: The Longest Revolution* (London: Virago, 1984), p. 299.

forms, although Freud concentrated most attention on 'conversion hysteria'. In the 'Little Hans' Case of 1909, he marks the difference between what were by then emerging as at least *two* forms of hysteria. In 'conversion hysteria' (in contrast to 'anxiety hysteria'), the patient 'converts' unresolved mental phenomena into bodily symptoms.[11] With conversion hysteria, the body of the hysteric speaks, and as its language is read so a new mode of interpretation is developed. The symptom is, in this sense, understood as a symbol or external sign of something which is taking place underneath – the body of the hysteric literally 'speaks', with a body language which expresses symbolically something which cannot otherwise be spoken. Hysteria is then a form of symbolic system, a 'culture' focused on the body. The psychoanalytic understanding of conversion hysteria as a 'malady through representation'[12] can then been seen as suggesting other forms of psychoanalytic cultural reading.

But first we must outline how Freud traced the aetiology of the disease from its mid-nineteenth-century image as a disorder of the nerves to its twentieth-century status as a psychological condition, in a way which facilitated the discovery of the basic principles of psychoanalysis – the analytic method itself, the existence of the unconscious, repression, fantasy, infantile sexuality. Feminists working with psychoanalytic theory have often pointed out that one of the first steps forward Freud and Breuer made in their work leading to the path-breaking publication of *Studies on Hysteria* (1893–5) was that, instead of looking at women and diagnosing them according to their bodily signs (as Charcot had done), they allowed them to speak, and developed forms of listening to them, to the unconscious speaking through their *ways* of speaking (their linguistic *style*), or even listening to their ways of being silent.[13]

The embryonic form of pre-psychoanalysis which evolved in the early 1890s was, then, already predicated on a set of linguistic exchanges between doctor and patient, a contract of speaker and analytic listener as writer and reader of unconscious experience. The analysand was encouraged to speak anything which emerged in analysis without self-censorship (Freud called this the 'fundamental rule of psychoanalysis'[14]). Both

11 In contrast to conversion hysteria, 'anxiety hysteria' is focused around a phobia: 'In anxiety hysteria,' he writes in 'Little Hans', 'the libido which has been liberated from the pathogenic material by repression is not *converted* (that is, diverted from the mental sphere into a somatic innervation), but is set free in the shape of *anxiety*' ('Analysis of a Phobia in a Five-Year-Old Boy ("Little Hans")', PF vol. viii, p. 274).

12 J. Laplanche and J.-B. Pontalis, *The Language of Psycho-Analysis* (London: Hogarth Press, 1983), p. 195.

13 Jacqueline Rose discusses this in 'Femininity and Its Discontents', collected in *Sexuality in the Field of Vision* (London: Verso, 1986).

14 In 'A Short Account of Psychoanalysis' (1924 [1923]) Freud writes (referring to himself in the third person): '[Freud] pledged his patients to refrain from any conscious reflection and to abandon themselves, in a state of quiet concentration, to following the ideas which occurred to them spontaneously (involuntarily)' (PF vol. xv, p. 166).

the use of hypnosis and Freud's later development of free association paved the way to the discovery and understanding of the structure of the unconscious – the idea was to develop a way of allowing the analysand to speak which would facilitate breaking down her conscious resistance to ideas, images, phrases which might emerge from forgotten parts of her memory, 'skim[ming] off the surface of their consciousness', as he put it later. He would pledge 'his patients to refrain from any conscious reflection and to abandon themselves, in a state of quiet concentration, to following the ideas which occurred to them spontaneously (involuntarily)'.[15] By definition, what this produced was a fragmented 'text' (which – as we will see later – Steven Marcus was to read as part of a larger modernist emphasis on the power of broken narratives). This is Freud discussing his abandonment of symptom-led techniques in favour of free association, some years later in the 'Dora' case (1905 [1901]):

> I now let the patient himself choose the subject of the day's work, and in that way I start out from whatever surface his unconscious happens to be presenting to his notice at the moment. But on this plan everything that has to do with the clearing-up of a particular symptom emerges piecemeal, woven into various contexts, and distributed widely over separated periods of time.[16]

Fidelity to this uncensored openness offered one form of access to 'the Royal Road to the Unconscious': 'I decided to start from the assumption', writes Freud in *Studies*, 'that my patients knew everything that was of any pathogenetic significance and that it was only a question of obliging them to communicate it.'[17] The term 'psychoanalysis' was then coined by Freud in 1896 to describe this very specific procedure of free association, through which the analyst can gain access to the unconscious parts of his patient's mind which are not accessible in normal conscious life.

But it is not the case that what patient Anna O. (in *Studies*) famously termed her 'talking cure' is simply a process of offloading readily interpretable material, even once the blocks, avoidances, and self-censorship involved in the process of telling have been successfully negotiated. These resistances in the patient's discourse were also crucial to Freud's developing understanding of the processes of repression – if the patient was now resisting bringing her traumatic narrative to the surface, this was only because at the moment of that trauma it was successfully pushed 'below'. Genuine psychoanalysis, Freud was to maintain much later in 1919, is the successful removal of amnesia.[18] Bringing *back* what was repressed through analytic working-through would finally *reverse*

15 Ibid.
16 Freud, 'Fragment of an Analysis of a Case of Hysteria ("Dora")', PF vol. viii, p. 41.
17 Freud in Freud and Joseph Breuer, *Studies on Hysteria* (1893–5), PF vol. iii, p. 173.
18 Freud, 'A Child Is Being Beaten', PF vol. x, p. 168.

the act of locking-up which the mind had enacted earlier in life.[19] This enables the analysand to reclaim and control subjective territory which had formerly controlled *her*, and this therapeutic process was famously identified by Freud in his dictum, 'where id was, there should ego come to be'. The ego, through analysis, would come to an understanding and mastery of the unconscious impulses which had 'undone' it in illness. If the unconscious, then, is an agent which 'decentres' the ego's sense of self and control, analytic therapy is partly about the humanistic practice of 'recentring' the ego. In identifying the governing primacy of the unconscious, psychoanalysis is 'antihumanist', but when it is engaged in the therapeutic practice of facilitating the analysand's recovery of personal control (the ability to love and to work, in Freud's definition of psychic health), it is practically humanistic. Two branches of post-Freudianism have taken these different strands forward: ego-psychology has developed the latter, narrowly therapeutic practice, and fixed upon a philosophy of the mastering ego, whilst Lacanian psychoanalysis has focused on exactly the opposite of this, reclaiming the 'decentring' power of the self-alienating Freudian unconscious (I shall return to this in Chapter 2).[20]

But, to return to the analysand's discourse, what would come out in analysis would not be transparent, and the process itself would be characterized by interpretation and resistance, bound by the terms of a developing transference relationship between analysand and analyst. The patient, in other words, does not offer up to the analyst an open 'text' thick with overt significance, but rather a linguistic and symbolic puzzle or jumble, which requires the training of a peculiarly 'literary' eye to read it powerfully. Forever sceptical about how subjects construct their narratives consciously (in the 'Rat Man' case Freud likens this to 'the process by which a nation constructs legends about its early history', saying what it *wants* to be true rather than what *is* true), Freud strives to read the analysand's discourse deviously, against the grain.[21] Whatever manages

19 'The impressions and mental impulses', Freud writes in 'A Short Account', 'for which the symptoms were now serving as substitutes, had not been forgotten without reason . . . they had, through the influence of other mental forces, met with a repression the success and evidence of which was precisely their being debarred from consciousness and excluded from memory' (ibid., p. 167).

20 This is discussed by Juliet Mitchell in 'Psychoanalysis: A Humanist Humanity or a Linguistic Science?', pp. 233–47 of *Women: The Longest Revolution*.

21 See Freud, 'Notes Upon a Case of Obsessional Neurosis (The "Rat Man")': 'It will help to put us upon the right track in interpreting it, if we recognize that more than one version of the scene (each often differing greatly from the other) may be detected in the patient's unconscious phantasies. If we do not wish to go astray in our judgement of their historical reality, we must above all bear in mind that people's "childhood memories" are only consolidated at a later period, usually at the age of puberty; and that this involves a complicated process of remodelling' (PF vol. ix, pp. 86–7n). I will look more closely at the processes involved in analytic narrative, as well as how psychic understanding takes place in stages and retrospectively, in Chapter 4 below.

to leak through in analysis can only be deciphered with difficulty and imagination, through a process of interpretation which suggests that right at the start of Freud's work, a developing intersection with literary analysis was taking place: just as psychoanalysis was coming into being, it was coming into being as a form of *active interpretation* of the patient's discourse *as text*. Freudian analysis thus takes on a crucial interpretative role – an analysis performed on and through the analyst's and the analysand's exchange of words begins to resemble nothing if not literary interpretation. As Jacques Lacan simply put it, the psychoanalyst reads 'dreams, slips of the tongue and even jokes *as one deciphers a coded message*'. Indeed, in 'A Short Account' Freud calls this early developing method 'the *art* of interpretation':

> When the 'fundamental rule of psychoanalysis' which has just been stated was obeyed, the course of free association produced a plentiful store of ideas which could put one on the track of what the patient had forgotten. To be sure, this material did not bring up what had actually been forgotten, but it brought up such plain and numerous hints at it that, with the help of a certain amount of supplementing and interpreting, the doctor was able to guess (to reconstruct) the forgotten material from it.[22]

Hysterics, then, carry with them moments from the past which are never quite present, but also never quite gone. Now, whilst hysteria is a very specific condition, Freud also comes to see it as an exemplary neurosis, produced in adult life by the symbolic expression of repressed childhood experiences (the Victorian hysteric thus becomes only a specific (if extreme) image of repressed 'normality').[23] Echoing Wordsworth's 'the child is father to the man', Freud writes in 'An Outline of Psychoanalysis' (1940 [1938]), 'The child is psychologically father to the adult'; 'the events of his first years are of paramount importance for his whole later life';[24] again, in *The Introductory Lectures on Psychoanalysis* (written 1915–17) he was to maintain that 'what is unconscious in mental life is also what is infantile'.[25] However, as early as *Studies* Freud was already beginning to think of the unresolved past and the unhappy present of the self as coexistent, bound together by a fault-line which prevented the individual ego from acting fully coherently. The past, then, is unsuccessfully held back, and engages in a constant invasion of the present through hysterical and other neurotic symptoms; as Breuer and Freud put it famously in *Studies*,

22 pp. 166–7.

23 '[I]n their hypnoid states they are insane, as we all are in dreams,' write Breuer and Freud, continuing, 'Whereas, however, our dream-psychoses have no effect upon our waking state, the products of hypnoid states intrude into waking life in the form of hysterical symptoms' (*Studies*, p. 64).

24 PF vol. xv, p. 421.

25 PF vol. i, p. 247.

'*Hysterics suffer mainly from reminiscences*'.[26] These are memories which have not been allowed to fade or be fully assimilated – traumatic memories charged with 'affect', which could not be discharged in the normal ways one might employ in order to get over a trauma (crying, taking revenge, 'blow[ing] off steam',[27] all common forms of 'reaction' which can discharge the affect attached to a memory): 'If reaction is repressed, the affect remains attached to the memory'[28] – for the hysteric, the affect remains 'strangulated'. Here Freud and Breuer begin to discuss the powers of language as an agent of catharsis – the ability of words to discharge the affect of trauma, or as Anna O. put it, the procedure of psychological 'chimney sweeping'.[29] The first use of the key terms 'abreaction' and 'catharsis' come in this passage, early in the first section of *Studies*:

> The injured person's reaction to the trauma only exercises a completely 'cathartic' effect if it is an *adequate* reaction – as, for instance, revenge. But language serves as a substitute for action; by its help, an affect can be 'abreacted' almost as effectively.[30]

Under hypnosis, then, Freud's patients revealed memories, the 'forgotten material' he describes much later in 'A Short Account', which '*correspond to traumas that have not been sufficiently abreacted*'.[31] Already an image of the mind as split and as potentially operating simultaneously on different time-scales is emerging. For the hysteric,

> ideas which have become pathological have persisted with such freshness and affective strength because they have been denied the normal wearing-away processes by means of abreaction and reproduction in states of uninhibited association.[32]

The memory – a past event – remains powerfully 'fresh', and time does not alter its affect – indeed, the past is *not* past until it is abreacted. Thus the hysteric (and, Freud is later to state) all 'normal' neurotics, live in a split state, divided by their ability to live and feel a number of past moments with the pain of a present trauma; it is here that he first talks of a 'splitting of consciousness'. It is here too (as early as 1892, when 'On the Psychical Mechanism of Hysterical Phenomena' was written) that psychoanalysis as a therapeutic practice begins to form itself (although

26 *Studies*, p. 58; their italics.
27 Ibid., pp. 58-9; this is part of the important opening section of the text, 'On the Psychical Mechanism of Hysterical Phenomena: Preliminary Communication' (1893), which traces the genesis of the hysterical symptom and gives a clear account of abreaction.
28 Ibid., p. 59.
29 Ibid., p. 83.
30 Ibid; their italics.
31 Ibid., p. 60; their italics.
32 Ibid., p. 62; their italics.

major changes are to come). In the much later essay 'The Question of Lay Analysis', Freud writes that '[t]he correct reconstruction . . . of such forgotten experiences of childhood always has a great therapeutic effect'.[33] The emphasis is then on piecing something together verbally and in terms of a sequence of images, reconstructing the repressed as a way of offering it a path through into consciousness, or, to put it the other way around (as Freud does in 1896), 'lead[ing] the patient's attention back from his symptom to the scene in which and through which the symptom arose'.[34] The emphasis is, then, on tracing the surface symptom present now back to its internal and past 'cause' – no longer a wandering womb, but a buried unconscious memory of trauma. Even in the 1892 paper, Freud and Breuer were already moving towards this:

> It will now be understood how it is that the psychotherapeutic procedure which we have described in these pages has a curative effect. *It brings to an end the operative force of the idea which was not abreacted in the first instance, by allowing its strangulated affect to find a way out through speech; and it subjects it to associative correction by introducing it into normal consciousness (under light hypnosis).*[35]

II

The Abandonment of the Seduction Theory and the Reality of Fantasy

Hysteria, then, was beginning to be understood as a condition characterized by the physical expression of something with a wholly psychosexual origin – the hysterical body 'spoke' the festering trauma which was at work, as it were, 'underneath': 'psychical conflict is expressed symbolically in somatic symptoms'.[36] Elements of a kind of linguistic exchange are already present in this very form of analysis: the body gives out signs which are a symbolic translation of past trauma replayed through the body in the present, and the analyst responds by turning body language into speech and actively interpreting the condition's cause, its originary trauma. The neurotic self is then built upon the uneasy foundation of the fault-line which sexuality represents, as prime site of childhood trauma

33 PF vol. xv, p. 317.

34 Freud, 'The Aetiology of Hysteria', trans. James Strachey, included as an appendix to *The Seduction of Truth* by J. M. Masson (Harmondsworth: Penguin, 1985), pp. 259–90, quotation p. 261.

35 'On the Psychichal Mechanism . . .', in *Studies*, p. 68; their italics.

36 Laplanche and Pontalis, 'Hysteria', in *The Language of Psycho-Analysis*, p. 194.

and anxiety: as Freud would later write, 'the weak point in the ego's organization seems to lie in its attitude to the sexual function'.[37] However, at this early point in his work, the 'forgotten material' he discusses is of a very specific – and real – kind. 'The memories which emerge, or can be aroused, in hysterica attacks', he writes again with Breuer,

> correspond to the precipitating causes which we have found at the root of *chronic* hysterical symptoms. Like these latter causes, the memories underlying hysterical attacks relate to psychical traumas which have not been disposed of by abreaction or by associative thought-activity.[38]

What, then, are these memories – what is it that emerges as 'of pathogenic significance'? As these embryonic analyses proceeded, Freud again and again heard accounts, emerging from the depths of repression, of how his patients had been 'seduced' (or sexually abused) in early childhood – by their fathers, by an older child, or by another adult. The clearest formulation of this position comes in two papers written in 1896, 'Further Remarks on the Neuro-Psychoses of Defence', and a paper delivered in 1896 to the Society for Psychiatry and Neurology in Vienna, 'The Aetiology of Hysteria', where he writes conclusively that '[w]hatever case and whatever symptom we take as our point of departure, *in the end we infallibly come to the field of sexual experience*'.[39] This, then, is the Seduction Theory (sometimes known as the 'Trauma Theory'), important because, firstly, it explicitly sexualizes children, and because it suggests that their sexuality is enforced by the act of another upon them, so disturbing and brutal that it is repressed: 'at the bottom of every case of hysteria there are *one or more occurrences of premature sexual experience*'.[40]

I discussed hysterical symptoms above as the return of repressed memories in a bodily form. In 'The Aetiology of Hysteria' Freud writes that 'hysterical symptoms are derivatives of memories which are operating unconsciously',[41] and at this point he is clear that what returns from the hysteric's repressed is actually predicated on a real event. What returns, then, is the repressed unconscious material of an unassimilated sexuality, and it is this which Freud and Breuer had begun to piece together in *Studies on Hysteria* as founding the physical symptoms from which their hysterical patients were overtly suffering. However, almost as soon as this is crystallized in 'The Aetiology of Hysteria', the position is questioned and then abandoned. The two tenets of the psychoanalytic

37 Freud, 'An Outline of Psychoanalysis', p. 421.
38 *Studies*, p. 66; their italics.
39 p. 267; his italics.
40 Ibid., p. 271; his italics.
41 Ibid., p. 280.

construction of the mind – the unconscious and the Oedipus complex/castration complex axis – are built upon an about-turn which Freud was to make a year later, upon which the model of the unconscious he was to continue working with for forty years was based. Freud's key epistemological break comes with a letter, written on 21 September 1897, to Wilhelm Fliess. 'I no longer believe in my *neurotica*,' he writes, giving four reasons (including a general inability to believe that all neurotics were the subjects of sexual abuse), most important of which is a crucial shift towards the primacy of fantasy: 'there are no indications of reality in the unconscious, so that one cannot distinguish between truth and fiction that has been cathected with affect.'[42]

The fantasy-life of children is, then, a 'fiction . . . cathected with affect' from which adults and their narratives will grow. This is not to say that children do not experience their fantasy-lives as truly traumatic, or that these lives are completely unattached from the reality of the family group – quite the opposite. To put it simply, instead of reading the hysteric's fantasies as direct reflections of an actual rape *by* their fathers, Freud began to see them as symptoms of what he was to call their Oedipal desire *for* their fathers – a taboo desire which, as well as encouraging the repressions he was already encountering the consequences of in later life, set the girls in a direct relationship of rivalry with their mothers. Hysteria is then grounded, as Freud succinctly puts it in the much later 'Short Account of Psychoanalysis' (1924 [1923]), in a repression based in a 'conflict between two groups of mental trends': 'Thus the symptoms were a substitute for forbidden satisfaction'.[43] Writing of his work in the third person, Freud sums up the move succinctly in 1922:

> The analytic researches carried out by the writer fell, to begin with, into the error of greatly overestimating the importance of *seduction* as a source of sexual manifestations in children and as a root for the formation of neurotic symptoms. This misapprehension was corrected when it became possible to appreciate the extraordinarily large part played in the mental life of neurotics by the activities of *fantasy*, which clearly carried more weight in neurosis than did external reality.[44]

The move was nothing if not controversial, and arguments have raged against this reversal at regular intervals ever since. In part, Freud has been accused of an appalling patriarchal defence of ostensibly blameless fathers over perverse, Oedipal daughters – he simply *could not* believe that all of the women who said they had been 'seduced' by their fathers

42 Freud to Fleiss, in *The Complete Letters of Sigmund Freud to Wilhelm Fliess 1887–1904*, trans. and ed. Jeffrey Moussaieff Masson (Cambridge, Mass., and London: Harvard University Press, 1985), p. 264.

43 'A Short Account', pp. 167–8.

44 Freud, 'Psychoanalysis', from 'Two Encyclopaedia Articles', PF vol. xv, pp. 141–2.

had in 'fact' experienced this.[45] Freud was, however, careful to address the existence of real sexual abuse in the histories of some patents; for instance, in the *Introductory Lectures on Psychoanalysis* he writes,

> A phantasy of being seduced when no seduction has occurred is usually employed by a child to screen the autoerotic period of his sexual activity. He spares himself shame about masturbation by retrospectively phantasying a desired object into these earliest times. You must not suppose, however, that sexual abuse of a child by its nearest male relatives belongs entirely to the realm of phantasy. Most analysts will have treated cases in which such events were real and could be unimpeachably established.[46]

However, what the abandoment of the seduction theory signals is a shift towards the primacy of *psychical reality*. A little earlier in the *Introductory Lectures* Freud writes, 'The phantasies possess a *psychical* as contrasted with material reality, and we gradually learn to understand that *in the world of the neuroses it is psychical reality which is the decisive kind*.'[47] With this, the question shifts from whether the memory of an event is always directly connected to such an event in empirical terms, or whether that 'memory' was generated in other ways, from infant desire, as a 'screen memory' which acts to conceal something else from consciousness, from sources of fantasy and real experiences already creatively worked on unconsciously by the patient (this is an issue which has taken on a new form in the recent 'false memory syndrome' debate). The universality of the 'memory' gave him pause for thought, too: that repressed memories of the child's sexualization so often take the same form is not because each analysand had *actually experienced* a *real* trauma inflicted upon them by an adult and always in the same way, but because the same 'themes' will occur across a variety of individual unconscious narratives: 'the sexual fantasy invariably seizes upon the *theme of the parents*'.[48] It is, indeed, the 'theme of the parents' which Freud was beginning to formulate as a universally-experienced set of desires and narratives which govern the development of gender and the formation of the unconscious – the Oedipus complex.

Another question was raised by something which emerged in Freud's own paternal make-up, which is pointed out by Juliet Mitchell in a brief

45 For a critique of Freud's abandoment of the seduction theory, see Kate Millett, 'Beyond Politics? Children and Sexuality', in *Pleasure and Danger: Exploring Female Sexuality*, ed. Carole S. Vance (Boston and London: RKP, 1984), pp. 217–24; Marie Balmary, *Psychoanalyzing Psychoanalysis: Freud and the Hidden Fault of the Father* (Baltimore, MD, and London: Johns Hopkins University Press, 1982), and Jeffrey Masson, *The Assault on Truth: Freud's Suppression of the Seduction Theory* (Harmondsworth: Penguin, 1985).

46 PF vol. i., p. 417.

47 Ibid., p. 415; his italics.

48 Freud, letter to Fliess, 21 September 1897, in *Complete Letters of Freud to Fliess*, pp. 264–5; my italics.

footnote in *Psychoanalysis and Feminism.* Here she quotes one of Freud's letters to Fliess, written prior to the letter of September 1897 quoted above, on 31 May. Here Freud reports to his friend a dream he has, charged with erotic feelings for his daughter, and interpreted by him as proving his own desire not primarily for the girl herself, but for direct proof that the cause of neurosis lies with perverted fathers, who in 'seducing' their daughters make hysterics of them: 'The dream of course shows the fulfilment of my wish to catch a Pater as the originator of neurosis and thus [the dream] puts an end to my ever-recurring doubts.'[49] Instead, it actually has the effect of reinforcing the doubts: if he *wanted* to 'catch a Pater as the originator', this may mean that the seduction theory is primarily the result of his own desire to blame the fathers, finding an easy answer in the real. In a culture which recognizes only empirically verifiable facts as true motivations for behaviour, any argument based on the existence of real traumas as the cause of all psychic discord would carry more weight than one emerging from a discussion of the rich fantasy life of all developing subjects. The presumed guilt of the father is then replaced in the theory by the repressed Oedipal fantasy-life of the child; the symptom becomes not an eruption of real seduction translated into the terms of the body, but 'a sign of, and a substitute for, an instinctual satisfaction which has remained in abeyance'.[50] 'Psychoanalysis', as Freud put it succinctly in 1925, 'disposed once and for all of the fairy-tale of an asexual childhood'; it lifts 'the veil of amnesia from the years of childhood'.[51]

So it is to the analysis of the individual's unconscious fantasy-life that Freud now turns, taking with him those readers also interested in the *other* fantasies we produce – literary, cinematic, representational narratives and images. Of particular weight for cultural criticism is the 'decentring' role of the unconscious and infantile sexual desires in adult life, and the way in which these are symbolically represented – as symptoms, dreams, parapraxes, even art-works as 'sublimated' desires. The Freudian unconscious is not so much a container full of repressed memories as a set of processes, processes which have a cultural resonance. In becoming repressed, and then in returning to consciousness in some form, repressed elements are subject to transformations enacted by the unconscious – displacement, symbolism, condensation, as meanings and the meanings attached to memories shift and are warped by the mind's constant reinterpretation of itself. In the *Three Essays on the Theory of Sexuality*, written seven years after the abandonment of the seduction theory, Freud returns to the hysterical symptom in an analysis of this

49 Ibid., p. 249. See also Juliet Mitchell, *Psychoanalysis and Feminism* (Harmondsworth: Penguin, 1974), p. 9.
50 Freud, 'Inhibitions, Symptoms and Anxiety', PF vol. x, p. 242.
51 'The Resistances to Psychoanalysis', PF vol. xv, pp. 271 and 272 respectively.

mental processing. The symptom is a substitute or 'transcription' of repressed wishes, underlying the fact that what is going on is already linguistic: repression entails a form of 'rewriting', but translated in this way, desire is inscribed on and below the surface of the body, rather than acted out in wishes.[52] The emphasis then shifts back to the inner lives of the hysterics and how they work on them, and with this shift the focus of analysis moves away from a search for empirically verifiable traumas buried in the self, to mapping out the landscape of fantasy. The analyst shifts from searching for buried evidence of a real event to working through exactly *how* the analysand responds to what happens to her, and how she speaks it. Her *style* of self-representation and self-betrayal becomes the focus.

Does this mean, then, that the patient is 'making it all up'? What would it mean to say this? This question already implies a pejorative view of fantasy. Since literary texts are also fantasies, how we judge something as real or unreal, fact or fiction, true lies or 'real' unconscious events, is clearly important. From 1897 Freud's position on this only gets stronger, as psychoanalysis increasingly emphasizes and credits the 'truth' of unconscious narratives – if it is real to the patient, it is real for psychoanalysis. As Juliet Mitchell puts it, underlining the broadly literary project which is unfolding here, 'Freud first believed that the stories were true and then that they were true as stories.'[53]

As readers or viewers we also believe that the texts we engage with are 'true *as stories*' (and no less 'true' for being 'stories'); so a psychoanalytic, anti-empirical suspension of disbelief could be said to be already taking place in a very untheorized way between reader and text, audience and cinema screen. In the shift towards an understanding of the self based primarily on unearthing its fantasy-landscape, Freud begins not only to offer ways of reading which will be useful in the analysis of literature but also to show that the self is itself made up of tales and images which are already 'literary' – imagistic and narrative fabrications woven from scraps of libidinally charged 'truth'. If we come into being through fantasy, we are at root its construction. We may be the stuff that dreams are made of, but more fundamentally we are made *by* those dreams.

An example of the way in which fantasies are credited as real in psychoanalysis is offered by the 'primal scene'. This is a crucial moment of acute anxiety and arousal in the early life of the child, in which an event – often an image of sexual intercourse between the parents – is witnessed

52 These repressed mental processes, Freud writes, 'being held back in a state of unconsciousness, strive to obtain an expression that shall be appropriate to their emotional importance – to obtain discharge; and in the case of hysteria they find such an expression (by means of the process of "conversion") in somatic phenomena, that is, in hysterical symptoms' (*Three Essays on the Theory of Sexuality*, in PF vol. vii, p. 78).
53 Mitchell, *Women: The Longest Revolution*, p. 299.

or imagined, or more usually imagined as having been witnessed, and then (as it is reworked and transcribed in memory) forms a turning-point in the development of that child as a sexual subject. Laplanche and Pontalis note that the German term *Urszenen* first appears in 1897 'to connote certain traumatic infantile experiences which are organised into scenarios or scenes'.[54] The way in which Freud develops his thought on the primal scene to some extent repeats his earlier shift away from real trauma towards traumatically experienced fantasy around the abandonment of the seduction theory, first discussing it as a scene of sex between parents *actually witnessed* by infants, then understanding it almost entirely as the most important 'primal phantasy'. Three texts written during the war years consolidate this most clearly: in *The Introductory Lectures on Psychoanalysis* Freud writes that 'Among the occurrences which recur again and again in the youthful history of neurotics . . . there are a few of particular importance . . . observation of parental intercourse, seduction by an adult and threat of being castrated',[55] although it is in two case histories that the first of these is really opened up. In 'A Case of Paranoia Running Counter to the Theory of the Disease' (1915) Freud says the same thing, adding, 'I call such phantasies – of the observation of sexual intercourse between the parents, of seduction, of castration, and others – "primal phantasies"'.[56]

The term is consolidated, however, in the 'Wolf Man' case (1918 [1914]), in which the analysand's neurosis is bound up with his image of 'a coitus *a tergo* [from behind], three times repeated': 'he was able to see his mother's genitals as well as his father's organ; and he understood the process as well as its significance'.[57] Freud is careful to stress the visibility of the act, but he also debated its reality – whether it is an imagined event, the consolidation of primal fantasy and desire. French psychoanalyst Jean Laplanche has developed the question of whether the scene is 'really' seen both in his work on fantasy with J.-B. Pontalis and in more recent discussions of *nachträglichkeit*, or deferred action, through which the significance of the primal scene is only opened up when it is encountered again and reinterpreted by the child later in life. Freud admits to this, again, in the *Introductory Lecture* discussion, where witnessing an act of sex between parents is only 'understood and react[ed] to . . . in retrospect'.[58] The child sees or hears *something*, but the material is itself only gradually inserted into a narrative or a coherent picture as it is actively reworked in memory – a reinterpretation and reinscription of the scene, taking place over time

54 *The Language of Psycho-Analysis*, p. 335.
55 Freud, *Introductory Lectures*, PF vol. i, p. 416.
56 Freud, 'A Case of Paranoia Running Counter to the Psychoanalytic Theory of the Disease', in PF vol. x, p. 154.
57 Freud, 'From the History of an Infantile Neurosis (The "Wolf Man")', in PF vol. ix, p. 269. I shall discuss this in more detail in Chapter 3.
58 pp. 416–17. This will be discussed in much greater detail in Chapter 4.

in the development of the subject, which is subsequently laced with further impressions coming from later life:

> If, however, the intercourse is described with the most minute details, which would be difficult to observe, or if, as happens most frequently, it turns out to have been intercourse from behind, *more ferarum* [in the manner of animals], there can be no remaining doubt that the phantasy is based on an observation of intercourse between animals (such as dogs) and that its motive was the child's unsatisfied scopophilia during puberty. The extreme achievement on these lines is a phantasy of observing parental intercourse while one is still an unborn baby in the womb.[59]

Fantasy, then, is not wholly fabricated, but it is an active (creative) construction of an image or fiction never actually witnessed, a patchwork of impressions and desires woven through with direct experience. Upon fantasy the subject is built (just as, Freud frequently points out, civilization is built upon myth). Unconscious fantasy is neither real nor unreal, except that it has a paramount reality in the life and construction of the subject. These are questions I shall return to later in subsequent chapters on fantasy and on the time and past of the subject. For the moment, it is important to grasp the point that, for all its ostensible visible veracity, the reality of the primal scene is contentious, as – in psychoanalysis at least – one sees what one has never laid eyes on, just as one may be blind to a sight straight ahead.

It is not, then, that the analysand makes up the stories of her early life, but that the stories have *made up her*. The subject is a creation *of* the story. The self comes into being in terms of, and entirely with reference to, the network of relationships which she later tells back to herself (as well as her analyst, listener and reader) as primal scenes, Oedipal crises, the narrative of early psychic history, stories which change as we change them, as we rework them through our individual histories. These, then, might actually be what Freud calls in one of the Fliess letters 'scientific fairy tales', narrations of the structures which construct the self, memories which are at root fantasies, and which have the telling scars of constant reworking, editing, and transcription about them.

III

The Case History as Story: *Dora*

Early psychoanalysis is then engaged in mapping out (as well as trying to resolve) a number of conflicts between speech and silence, expression

59 Ibid., p. 417; Freud's italics, editor's interpolation.

and muteness, with the hysterical body at its origin, 'representing' malady in the wake of a failure of conventional language. Charcot 'reads' the hysteric's symptoms; Freud gets her to speak – of dreams, memories, fantasies – and interprets the words (often against the grain) instead. In its basic processes psychoanalysis is, then, already predicated upon a 'literary' exchange, with the unconscious as text-to-be-read. As Freud's work evolves, particularly in the areas of dream analysis, the interpretation of parapraxis (slips of the tongue or the pen), and the development of an ever more sophisticated system for analysing utterance, so the focus becomes *how* one says what one says, how one's form of speech might contradict or subvert the tale one is telling, rather than what one 'means' to say consciously and with full moral purpose. In stressing the forms of speech and imagery which emerge in analysis, the literary-linguistic nature of both the analytic situation and the divided mind itself becomes ever clearer. It is not simply that Freud is moving psychoanalysis towards identifying that language is instrinsic to our understanding of mental processes (which the French psychoanalyst Jacques Lacan was to develop later), but that psychoanalysis is becoming increasingly sensitive to the complexities of the patient's words, the way in which words betray desires. If literary language is of a very specific type – allowing for ambiguities of meaning, imagery which may point in several directions, complex symbolic systems and possible interpretations laid down entirely against the grain of what the speaker or writer might overtly or consciously mean – then so is the language of analysis, and in many ways the processes involved in understanding both are the same.

I want now to pursue some of these issues by turning back to the hysteric in her literary form. She has continued to be something of a feminist, as well as a literary, heroine, so it is not only because these early psychoanalytic concerns are so self-consciously linguistic that hysteria is an important focus for contemporary literary studies. Hysteria has been crucial to debates on and in the subject, whether it is read for its important historical role in developing psychoanalysis as a form of linguistic exchange and interpretation, as a form of feminist protest (if the woman can't speak, then her body *will*), or as the key to a literary enigma (a number of Victorian heroines have been read in terms of discussions of hysteria, ranging from Edmund Wilson on the governess of *The Turn of the Screw*, to Mary Jacobus on Lucy Snowe in *Villette*, or Jacqueline Rose on George Eliot's Gwendolen Harleth[60]).

This work on narrative and linguistic construction also has more literal

60 See Wilson, 'The Ambiguity of Henry James', in *The Triple Thinkers* (Harmondsworth: Penguin, 1962); Jacobus, 'The Buried Letter: Feminism and Romanticism in *Villette*', in *Women Writing and Writing About Women*, ed. Jacobus (London: Croom Helm, 1979), pp. 42–60; and Rose, 'George Eliot and the Spectacle of Woman', in *Sexuality in the Field of Vision*, pp. 104–22.

implications. Freud's case histories have been increasingly read as short stories, particularly since Steven Marcus's pathbreaking analysis of the 'Dora' case as literature in 1974 (to which I shall turn later), read – in Peter Brooks's terms – 'for the plot', and constructed around certain conventions of what makes a good tale. Anna O.'s 'talking cure' enacts a story which is a reworked repetition of the events which first brought her symptoms into being – in telling the story again (in her own words, as it were), she releases the hold of its old formulation.[61] In discussing the case of Fräulein Elisabeth Von R. in *Studies on Hysteria*, Freud makes explicit the connection between the patient's narrative and a work of fiction:

> I have not always been a psychotherapist. Like other neuropathologists, I was trained to employ local diagnoses and electro-prognosis, and it still strikes me as strange that the case histories I write should read like short stories and that, as one might say, they lack the serious stamp of science. I must console myself with the reflection that the nature of the subject is evidently responsible for this, rather than any preference of my own. . . . Case histories of this kind are intended to be judged like psychiatric ones; they have, however, one advantage over the latter, namely an intimate connection between the story of the patient's sufferings and the symptoms of his illness – a connection for which we still search in vain in the biographies of other psychoses.[62]

The 'story' is then inextricably bound to the symptom. It is hysteria itself which is intrinsically 'literary' in origin and cure, drawing the dispassionate scientist away from his training into the world of narrative, for cure here depends upon a particular form of story-telling. Freud's work was increasingly to 'lack the serious stamp of science', relying more and more on reading processes, especially in the six major case histories published well after this statement, from 1905 to 1920, only one of which concerned conversion hysteria.

His own 'preferences' were to change and develop too, as the literary metaphors at the heart of the psychoanalytic structure were increasingly

61 Anna O's symptoms were, of course, already specifically linguistic. Charles Bernheimer discusses Dianne Hunter's work on this in his introduction to *In Dora's Case* (pp. 8–9). Anna O 'translated her conversion symptoms into the narrative of their origin, thereby undoing them. This translation, as Dianne Hunter has recently stressed, involved the movement not only from body language to verbal expression but also from multilingualism to a single language. For among her symptoms had been an inability to understand or communicate in her native German and a tendency to speak in one or more foreign tongues, in sequence or, at times of extreme anxiety, in an unintelligible mixture. This disruptive polylingualism, Hunter argues, may reflect a refusal of the cultural identity inscribed in the order of (coherent) German discourse and an unconscious desire, become conscious in certain contemporary feminist writers, to explode linguistic conventions'. See also Hunter, 'Hysteria, Psychoanalysis, and Feminism: The Case of Anna O', *Feminist Studies*, 9/3 (Fall 1983).

62 p. 231.

foregrounded. The model for the Oedipus complex itself comes, of course, from a literary text (Sophocles' drama *Oedipus Rex*). 'Mother-incest was one of the crimes of Oedipus, parricide was the other,' Freud writes in the *Introductory Lectures*, naming the taboo desires bound up in the complex which, initially, was the model through which the infantile sexuality of both girls and boys was understood (the girl's desire, sometimes known as the 'Electra complex', was equally heterosexual in Freud's early models, with the girl desiring the father and rivalling the mother in a mirror-image of the boy's situation). Freud continues in the *Introductory Lectures* in a literary vein:

> What help does analysis give towards a further knowledge of the Oedipus complex? That can be answered in a word. Analysis confirms all that the legend describes. It shows that each of these neurotics has himself been an Oedipus or, what comes to the same thing, has, as a reaction to the complex, become a Hamlet.[63]

This is only one amongst a number of scattered references to Shakespeare in Freud's corpus, in which he reads the equally Oedipal *Hamlet* as a modern version of the earlier myth (Hamlet is Oedipus with the added factor of repression).[64]

What is at stake here, however, are the processes which underpin analysis itself, and further issues are raised by Freud's 1905 case, 'Fragment of an Analysis of a Case of Hysteria ("Dora")' (I will refer to the case itself as 'Dora' to distinguish it from its subject, the analysand herself). If case histories are short stories, 'Dora' lacks the closure of a complete tale, and is perhaps the Freud text read most frequently against the grain in terms of its own gaps, omissions, disavowals – there has been an exciting history of readings and psychoanalytic challenges to Freud through the 'Dora' text, both through Freud's own role as unreliable narrator and taking the text itself in other ways as the subject of analysis. Dora (whose real name was Ida Bauer) was taken to Freud by her father as an eighteen-year-old girl, suffering from a range of hysterical symptoms and threatening suicide. She and her parents had recently spent a holiday with a couple, Herr and Frau K., during which time Dora accused Herr K. of propositioning her, which K. counters by charging Dora with making it all up. In analysis, Freud uncovers an older story, in which Dora's father has long been engaged in an illicit affair with Frau K. (who is also friend to Dora, and something of a maternal figure), and Dora is the bargaining chip given to Herr K. as a trade-off. The situation

63 pp. 378–9.

64 Freud's most substantial readings of *Hamlet* as a literary case history are in the section on 'Dreams of the Death of Persons of Whom the Dreamer Is Fond', in *The Interpretation of Dreams*, (PF vol iv, pp. 366–8), and the 1905–6 essay, 'Psychopathic Characters on the Stage' (collected in PF vol. xiv).

had persisted since Dora was fourteen, but by the time she is eighteen she has had enough. Whilst recognizing the sordid inter-family web in which Dora is impossibly positioned, Freud nevertheless persists in reading her hysterical symptoms through two detailed and lengthy dream analyses[65] in terms of repressed Oedipal desire and jealousy of Frau K., in the position of the mother:

> According to a rule which I had found confirmed over and over again by experience, though I had not yet ventured to erect it into a general principle, a symptom signifies the representation – the realization – of a phantasy with a sexual content, that is to say, it signifies a sexual situation.[66]

Refusing this account, Dora leaves the analysis and the case remains (as far as Freud is concerned, at least) an open one – essentially a textual fragment, unfinished.

In its gaps and blindnesses to the factors which might contradict or complicate this account, the 'Dora' case has offered a rich source for debate ever since. Here in particular, the process of transference is foregrounded, another dimension to the question of analytic reading. What the patient 'reads into' her analyst (transference – the displacement of feelings attached, say, to parents onto the analyst in order to work them through) is overtly an issue for Freud. Transference emerges through the development of Freud's work as a kind of acting-out, an active remembering, through analysis, in which the patient transfers the key components and basic dynamic of her fantasies and memories onto the screen of her relationship with her analyst:

> the patient does not remember anything of what he has forgotten and repressed, but acts it out. He reproduces it not as a memory but as an action; he repeats it, without, of course, knowing that he is repeating it.[67]

Transference, in this formulation, is then a form of play-acting; if the symptom is pantomime, so its transferential 'staging' acts out the cure. The analyst is also part of the play, however, and what he, in turn, 'reads into' the analysand's narrative (through counter-transference) comes to the fore later, particularly in Jacques Lacan's 1951 essay, 'Intervention on

65 The 'Dora' case was to be called 'Dreams and Hysteria': 'the explanations are grouped around two dreams; so it is really a continuation of the dream book' (letter to Fliess, 25 January 1901, in *Complete Letters of Freud to Fliess*, p. 433). In the case itself, Freud writes: 'I wished to supplement my book on the interpretation of dreams by showing how an *art*, which would otherwise be useless, can be turned to account for the discovery of the hidden and repressed parts of mental life' (PF vol. viii, p. 155; my italics).

66 Ibid., p. 80.

67 Freud, 'Remembering, Repeating, Working-Through', in *Standard Edition of the Complete Psychological Works of Sigmund Freud* (London: Hogarth Press, 1955–74; hereafter *SE*), vol. xii, p. 150.

Transference'. Just as we bring to texts a whole range of cultural and sexual baggage, so analysis does not take place in a desire-vacuum; there are, as Peter Brooks put it, 'real investments of desire [coming] from both sides of the dialogue'.[68] Freud reads Dora's turbulent relationship with himself partly as a mis-handling of the 'normal' transference of the past patternings of Oedipal desire between herself and her father/Herr K. displaced onto the present relationship of analyst and analysand. The phenomenon of patients falling in love with their analysts had been known since Breuer's intense relationship with Anna O., and Freud rationalizes such situations as a replay, through analysis, of the key relationships the analysand is still desperately trying to resolve. Transference is thus crucial to the successful course of analysis, since it facilitates the patient's reworking of her unconscious story.

But if Dora reads into the current situation a past moment, so too does her analyst. The case also formed the site upon which the notion of counter-transference is most clearly developed, specifically through the influence of Lacan's reading (the few moments at which Freud explicitly addresses the subject concentrate on the dangers of the analyst's desire governing and guiding the analytic relationship). For Lacan, '*psychoanalysis is a dialectical experience*'[69], and the 'Dora' case is predicated upon a series of reversals through which desire and the interpretation of roles are channelled at different times in *both* directions, from Dora to Freud and back again in an exchange not just of desire but of identification. Ever the literary critic regarding analytic situations, Lacan reads against the grain, and knowingly charges us to obey that old practical critical rule: 'As in any valid interpretation, we need only stick to the text in order to understand it.'[70] By sticking to the text, Lacan reads Freud counter-transferentially as positioning himself in the role of Herr K. regarding Dora, putting himself 'rather too much in the place of Herr K.' So Dora leaves Freud as she has left K., 'withdraw[ing] with the smile of the *Mona Lisa*'.[71] Freud also brings to his analysis a range of assumptions about Dora's desire (for men, for him) which blinded him to other currents at work in the scene. It is, then, Freud's (counter-transferential) desire which comes to dominate the case, not Dora's transference. Thus just as the analyst reads the gaps and omissions of the patient's narrative, so he also inserts and lays bare his own gaps in understanding and

68 'The Idea of a Psychoanalytic Literary Criticism', in *Discourse in Psychoanalysis and Literature*, ed. Shlomith Rimmon-Kenan (London and New York: Methuen, 1987), pp. 1–37, quotation pp. 12–13.

69 Lacan, 'Intervention on Transference', p. 63; his italics.

70 Ibid., p. 70.

71 Ibid.

discourse, producing significant 'failures' in reading which critics have in turn opened up in subsequent interpretations.[72]

One such productive 'failure' concerns the nature of narrative itself. If – as is said of the Victorian novel – all good stories end with a marriage, then it could be said that Freud's role as writer of psychological histories which 'read like short stories' places him in the position of narrator, under pressure to deliver a satisfying plot trajectory and a sense of closure as the tale reaches its conclusion. The analysand engages in symbolic, representational work in producing the symptom, which is then read through her free association and dream-account, and transcribed as the story of her case history, with beginning, middle, and (it is hoped) an end, infused with mystery and suspense animating each analytic discovery. The analyst, like the literary critic, is the writing-detective in the drama of the case.

The 'Dora' case has no such conclusion, however; as unfinished, annotated, edited and returned to by Freud (who added footnotes in 1923), it confounded the conservative patternings of conventional narrative. Two readings of 'Dora's' 'literariness' explore the sheer unwillingness of the case to conform to realist story-telling. In 'Enforcing Oedipus: Freud and Dora', Madelon Sprengnether discusses how the problem with Freud's resolution of (or rather his failure to resolve) Dora's story lies in his (counter-transferential) adherence to a plot structure which dictates the dominance of heterosexual closure and desire. The gap in the narrative, like the lack in Dora, must be filled; the case, writes Sprengnether,

> appears to be structured around a central irony – the attempt to complete a story and to achieve narrative closure rendered forever impossible through Dora's deliberate rupture. . . . Freud, in his pursuit of a phallic interpretation of Dora's desire, urging her toward a heterosexual pact in which her gap will be filled and his case history brought to a suitable conclusion, does not perceive the way in which phallic aggressiveness itself acts as a symptom.[73]

However, in his influential and energetic 1974 essay, 'Freud and Dora: Story, History, Case History' (one of a growing body of work which has opened up the reading of Freud as a literary figure[74]), Steven Marcus discusses the case as a modernist fable which is important precisely because it *fails* to tie up its own loose ends. If analysis is about the therapeutic

72 The Bernheimer/Kahane collection *In Dora's Case* offers an exemplary (though by now not exhaustive) range of readings, only some of which I am able to touch on here.

73 'Enforcing Oedipus: Freud and Dora', in *In Dora's Case*, pp. 254–75, quotation pp. 268–9.

74 Marcus writes at the opening of the essay, 'My assumption – and conclusion – is that Freud is a great writer and that one of his major case histories is a great work of literature' (*In Dora's Case*, pp. 56–91, quotation p. 57). This has been richly developed elsewhere; see texts by Hertz, Mahoney, Meisel (ed.), Skura, Smith and Trilling in the Bibliography at the end of this volume.

telling of stories, Marcus identifies a link in Freud's assumptions between narrative and mental health, or rather, between the hysteric's 'narrative insufficiency', her inability to 'tell a coherent story of their lives', and the condition from which she suffers.[75] Analysis, as it emerges in Marcus's reading of Freud's reading of Dora's story, is about getting one's story straight.[76]

But *whose* story is the 'Dora' case? Here, '[t]he patient does not merely provide the text; he also is the text, the writing to be read, the language to be interpreted',[77] but it is exactly this textuality which allows Dora to slip out of sight. Freud 'is as much a novelist as he is an analyst'[78] (and thus, to return to our earlier discussion of the reality of fantasy, he is 'making it all up' as much as any fantasizing patient ever is; Marcus writes that 'the "reality" Freud insists upon is very different from the "reality" that Dora is claiming and clinging to'[79]). The case can then be read as a kind of 'modern experimental novel', elliptical, 'at odds with itself', 'neither linear nor rectilinear'. As with fiction, so with analytic writing – the identifications here are multiple:

> In this Ibsen-like drama, Freud is not only Ibsen, the creator and playwright; he is also and directly one of the characters in the action and in the end suffers in a way that is comparable to the suffering of the others.[80]

As Marcus is to demonstrate, however, Freud is not just a character in the drama of his making, equivalent to any of the other active players. Rather, for reasons to do with the failure of the transference – or the failure of Freud to master his own counter-transference – he becomes the central character, 'like some familiar "unreliable narrator" in modernist fiction',[81] just as his heroine is refusing her role and exiting through the door. Because it is a case of counter-transference, the 'Dora' case is then more the story of Freud than the story of Dora. It is *Freud's* (flawed) reading of Dora which emerges as the case's central thread: 'Instead of letting Dora appropriate her own story, Freud became the appropriator of it',[82] and, callous though this sounds, it is precisely *because* of this therapeutic failure that the case itself becomes so interesting as literature (and so

75 Marcus, 'Freud and Dora', pp. 70–1.

76 'At the end – at the successful end – one has come into possession of one's own story. It is a final act of self-appropriation, the appropriation by oneself of one's own history. . . . What we end with, then, is a fictional construction that is at the same time satisfactory to us in the form of the truth, and as the form of the truth' (ibid., pp. 71–2).

77 Ibid., p. 81.

78 Ibid., p. 79.

79 Ibid., p. 85.

80 Ibid., pp. 64–5.

81 Ibid., p. 66.

82 Ibid., p. 85.

problematic for feminism). Freud admits that he 'did not succeed in mastering the transference in good time', but Marcus's point is that in writing up the case he worked through the situation (and continued to do so in his compulsive *return* to the scene as the case is revised and republished): 'this cluster of unanalyzed impulses and ambivalences was in part responsible for Freud's writing of this great text immediately after Dora left him. It was his way – and one way – of dealing with, mastering, expressing, and neutralizing such material.'[83] The failure of transference and counter-transference (Freud doesn't seem to like Dora very much) accounts for the success of the writing. It doesn't matter that the analysis failed; rather, its failure as therapy is an index of its success as a complex, tortured form of writing which is also incessantly engaged in reading itself. Freud thus becomes writer, reader and actor in his own problematic and ambivalent text: 'Freud's case histories are a new form of literature; they are creative narratives that include their own analysis and interpretation.'[84]

IV

The Story as Case History: *Cousin Phillis*

So if case histories are stories, are stories in turn case histories? This is the question which has underpinned much of psychoanalytic literary criticism this century, and indeed it is raised initially in the very procedure Freud adopts in the 'Schreber' case, one of the six major histories mentioned above, which is based entirely upon a textual reading.[85] Who or what is the subject of the literary case – the character (Hamlet, for instance, in those scattered readings I mentioned), the text itself, or the author? In the 'Schreber' case, the analysand is the writer himself, and is present only in his autobiography (*Memoirs of My Mental Illness*, by Daniel Paul Schreber), which Freud proceeds to read with great care, producing both an intricate textual reading and a diagnosis. Reading 'Dora' as literature is one thing; reading autobiography as free association (or its equivalent) is another. Reading Elizabeth Gaskell's *Cousin Phillis* for its account of hysteria, or its incipient Oedipality, opens up a number of further questions. In the last section of this chapter, I want to use this short novel as a way of addressing some basic questions which will set the terms for each of the chapters below.

83 Ibid., p. 90.

84 Ibid.

85 Freud, 'Psychoanalytic Notes on an Autobiographical Account of a Case of Paranoia (Dementia Paranoides) (Schreber)' (1911 [1910]), PF vol. ix.

The fictional hysterical woman has been one of the most rewarding figures for feminist readers, who have traced her role in literary (particularly classic realist) texts and women's history. As we have seen, hysteria was the first fully psychoanalytic disorder, in that it was Freud's work with hysterics in the 1890s which took him through the seduction theory to its abandonment, crucial to the discovery of the unconscious and the formulation of psychoanalysis proper. But if hysterical women guided Freud to a form of analysis which necessitated a theory of the unconscious and a non-empirical account of psychic development, in their readings of literary hysteria feminists have often moved toward rather more reason-based interpretations. Hysteria has been read as an eminently sensible response to impossible circumstances, an active refusal to accept intolerable conditions, or an unwillingness to choose between masculine and feminine identifications, love-objects, speech or silence. Discussions of Charlotte Perkins Gilman's story *The Yellow Wallpaper* offers an example of this way of reading, with the heroine's insanity read as a legitimate form of psychic escape carried out in response to the patriarchal oppression to which she is subject.[86] The hysterical heroine has then been equally valorized and mourned, a revolutionary figure protesting against her lot, by turns victim of, or in revolt against, the sexually conflicting or oppressive conditions which engendered her hysteria. Hélène Cixous, in discussion with Catherine Clément, reads Dora in this way:

> Dora seemed to me to be the one who resists the system, the one who cannot stand that the family and society are founded on the body of women, on bodies despised, rejected, bodies that are humiliating once they have been used. . . . It is the nuclear example of women's power to protest. It happened in 1899; it happens today wherever women have not been able to speak differently from Dora, but have spoken so effectively that it bursts the family to pieces.[87]

Does Elizabeth Gaskell's *Cousin Phillis* lend itself to this form of reading? What would it mean to engage in a diagnosis of hysteria played out at

86 Readings which discuss *The Yellow Wallpaper* diagnostically through its heroine's illness as rebellion or symptom of sexual/social repression include Gilbert and Gubar, Kennard, Kasmer, Kolodny (see Bibliography). Janice Haney-Peritz counters this by reading the tale as the 'story of John's demands and desires rather than something distinctively female' (Haney-Periz, 'Monumental Feminism and Literature's Ancestral House: Another Look at *The Yellow Wallpaper*', *Women's Studies*, 12 (1986), p. 123, also quoted by Kasmer, 'Charlotte Perkins Gilman's "The Yellow Wallpaper": A Symptomatic Reading', *Literature and Psychology*, 36/3 (1990), p. 13.

87 'The Untenable', in *The Newly Born Woman* by Cixous and Clément ((Manchester: Manchester University Press, 1986), p. 154. Cixous continues: 'The hysteric is, to my eyes, the typical woman in all her force. It is a force that was turned back against Dora, but, if the scene changes and if woman begins to speak in other ways, it would be a force capable of demolishing those structures.' Clément argues against this position: 'What [Dora] broke was strictly individual and limited' (p. 157).

the level of plot and theme on the page? *Cousin Phillis* is something of a Tardis-novel: slight, short, its issues provincial and small-scale – essentially the crisis of one obscure adolescent girl in a Cheshire backwater, whose story is rendered through an observance of changes in her body, her gestures, her move into physical language as she becomes increasingly mute and her breakdown is acted out. Within these narrative limits, the minute terrain of the text opens up into a series of increasingly vertiginous scenes of breakdown, loss and impossible choices focused on the heroine's body and the family's language. The story unfolds thus: Paul (our young first-person narrator) leaves home to work on the new railways which are making inroads into a corner of rural England, unchanged for centuries. He makes contact with his distant relatives, the Holman family, headed by Reverend Holman, who is a farmer as well as a holy man (indeed, his two roles combine to help him unite his immediate community in an organic, pre-industrial idyll. Paul feels as if he has wandered into the Old Testament, and, indeed, the changes which are to come through Phillis's growing-up and breaking-down are like a move into a new biblical era). Holman's seventeen-year-old daughter Phillis has never been away from her home, but Paul is slightly peeved that a girl two years his junior is rather taller and infinitely smarter than he: she has a keen intellectual thirst, reads avidly, and her father – with whom she identifies closely – has taught her Latin and Greek. Paul brings one of his colleagues, Holdsworth, to meet the family, who becomes ill and has to stay with them until he recovers, during which time he teaches Phillis Italian, and she in turn falls in love with him. Indeed, he exercises an 'unconscious hold' on all of the family ('It is like dram-drinking,' says Holman, 'I listen to him till I forget my duties, and am carried off my feet'[88]). Paul, who early on had entertained fantasies of courting Phillis himself, now becomes her confidante. Holdsworth goes away to Canada, and Phillis is deeply but secretly upset, so Paul reveals that Holdsworth had himself confessed his love for Phillis, and his hope to marry her. The whole episode has unbalanced Phillis (who has no experience of an emotional life beyond the bounds of the family), and when a letter arrives some time later, announcing Holdsworth's marriage to someone else, Phillis is pushed over the brink, and becomes gravely ill. The crisis reaches out to her father too, who has no conception that his daughter could love another, or could keep a secret from him. The novel closes with Phillis reaching a precarious state of recovery.

This is not the stuff of world-historic apocalypse, it is true. Instead, *Cousin Phillis* plays out a small-scale crisis taking place largely in the internal life of a young woman in an obscure corner of nineteenth-century England. Nevertheless, what animates this story is one of the 'grand

88 Elizabeth Gaskell, *Cousin Phillis* (Harmondsworth: Penguin, 1986), p. 266.

narratives' which psychoanalysis has identified most powerfully – a reversed version of a female Oedipal crisis, exploring closely the father's response to his daughter's crisis, played out through an intricate drama which breaks down the shared meanings of the family. If *Hamlet* is Oedipus with repression, and Lawrence's *Sons and Lovers* plays out the agonies of son caught between the desires of mother and lover, *Cousin Phillis*, haunted by the primary bond between daughter and father which underlies all other relationships in the text (she takes after him, he is her first love), plays out the daughter's break followed by the father's pain. The very force of Phillis's attraction for Holdsworth, the outsider, tears open in her an emotional impossibility – love for her father or for Holdsworth, the stranger; masculine identification *with* her father or the femininity which heterosexual attraction seems to enforce.

How does Freud set up this model of Oedipal development? The Oedipal moment occurs when the (male) child moves aways from pre-Oedipal unity with the mother, and begins to see himself in a relationship of conflict with his father based on his desire for his mother ('Mother-incest was one of the crimes of Oedipus,' Freud has said above in that quote from the *Introductory Lectures*; 'parricide was the other'). The pattern is three-cornered: the child sees the father as the intervening threat between himself and his mother, and he also sees his place between his father and mother as potentially drawing his father's wrath – he fears castration as a result of his father's anger at his incestuous desire (the boy-child 'successfully' moves from the Oedipus complex through to the castration complex). He then represses his desire for his mother – now perceived as taboo – and exchanges it for identification with the father, looking forward to a 'mature' heterosexual desire for other women when he reaches puberty. The most famous modern cultural representation of this, written in a full awareness of the developing ideas of Freud, is indeed *Sons and Lovers* (1913), which focuses on the dilemma of Paul Morel, split between his passionate attachment to his mother (for whom he has become a substitute husband) and his growing interest in other women. One of the earliest examples of psychoanalytic literary criticism was formed around this novel – a powerful (if by today's standards a 'crude Freudian') reading by Alfred Booth Kuttner in 1916 of Paul Morel as Oedipal image of the author.[89] In his difficult position between Oedipal love and fear of the father, the child is pinpointed

89 See Alfred Booth Kuttner, '*Sons and Lovers*: A Freudian Appreciation', originally published in *The Psychoanalytic Review* (July 1916), reprinted in '*Sons and Lovers*': *A Casebook*, ed. Gámini Salgádo, (London: Macmillan, 1975).

by an anxiety of triangulation.[90] Using this three-cornered set of family co-ordinates, the child uneasily orientates himself towards sexual identity.

As with boys, so with girls, only 'through the looking-glass' – in his early models of 'normal' feminine sexuality, heterosexual female development is the mirror image of the above model: it is the father the little girl loves, and the mother with whom she is in a relationship of rivalry. The girl lacks a penis (she is 'already' castrated), but she deals with her penis-envy by transforming it into the desire for a (the father's) baby, a mature form of heterosexuality predicated on an original love for the parent of the opposite sex. In his later models (for instance, in the 1932 essay 'Femininity'), Freud revises this in the light of the fact that girl-infants also start life in a relationship of pre-Oedipal unity with the mother, which means that the mother is the original object-choice for both girls and boys. These ideas will be expanded in various ways in subsequent chapters; particularly, the girl's pre-Oedipal attachment to the mother will be the focus for feminist psychoanalytic theories, discussed in Chapter 3.

The relevance of this to Cousin Phillis could be charted in quite crude terms of Phillis's relationship with her father (though perhaps less spoken of than any other relationship in the novel, it is the most important). Phillis's father places himself in a direct position of rivalry with Holdsworth, fighting against talk of '*another* man's love' for her.[91] But aspects of this family-patterning ripple across the text on various intricate levels. This is a novel which works in threes, as subjects are caught in a number of shifting triangulations, in space as well as desire. The subject is constructed with reference to the desires of (two) others. On the second page Paul finds himself in a room that 'was all corners, and everything was placed in a corner, the fire-place, the window, the cupboard; I myself seemed to be the only thing in the middle, and there was

90 As well as its connotation in orienteering, the term 'triangulation' has most cultural currency in René Girard's *Deceit, Desire and the Novel* (1961; Baltimore, MD, and London: Johns Hopkins University Press, 1969), which uses a Hegelian model to figure desire of subject for object as always mediated by a third term: 'The spatial metaphor which expresses this triple relationship is obviously the triangle', he writes, here of *Don Quixote*, but his textual analyses concern more 'permanent' emotional states: 'The object changes with each adventure but the triangle remains' (p. 2). Feelings of threat, for instance, are classically 'triangular': 'Jealousy and envy imply a third presence: object, subject, and a third person toward whom the jealousy or envy is directed. These two "vices" are therefore triangular' (p. 12). Later Girard writes: 'From a Freudian viewpoint, the original triangle of desire is, of course, the Oedipal triangle. The story of "mediated" desire is the story of this Oedipal desire, of its essential permanence beyond its ever changing objects' (pp. 186-7n).

91 *Cousin Phillis*, p. 307; my italics.

hardly room for me'.[92] Here the narrative leads directly into a discussion of familial triangulation which sets the tone for his (and our) reading of the situation in which Phillis is immersed. Paul is not just a human figure in a room full of corners, but a child caught within, and constantly reinterpreting, parental desire:

> I was an only child; and though my father's spoken maxim had been, 'Spare the rod and spoil the child,' yet, unconsciously, his heart had yearned after me, and his ways toward me were more tender than he knew, or would have approved of in himself could he have known. My mother, who never professed sternness, was far more severe than my father: perhaps my boyish faults annoyed her more; for I remember, now that I have written the above words, how she pleaded for me once in my riper years, when I had really offended against my father's sense of right.[93]

This is a passage which deals overtly with levels of knowing and behaviour, with unconscious relationships played out beneath, and simultaneously with, conscious ones. The sentence extends through a stream of sequential but only covertly connected thoughts, working through the boy's mediation of adult relationships, his varying terms of reference with mother and father being interpretable through a post-Oedipal model of identification and action. However, perhaps the most interesting turn here comes with the word 'for' after the semi-colon of the second sentence. This is a point of contradiction, where Paul turns back on what he has just said and re-remembers it in his mother's favour – one would, then, expect that word to be 'but' ('my boyish faults annoyed her more; *but* I remember, now I have written the above words, how she pleaded for me . . .'). As the sentence stands, the consequence of one statement from the other is not clear or easy. The work that 'for' is doing here is not the work of linear logic. The memory of the pleading mother ought not to run on from the memory of the stern mother, except if the second thought is already there and buried in the first (the difference between his 'boyish faults' and his 'riper years' notwithstanding). Two relationships with the mother are then being allowed here, coexisting (in 'for') rather than contradicting each other (as 'but' would imply). As Paul's flawed or layered memory unravels itself in the narrative, it presents him with, and allows, two contrary images (she was with him, and against him) to coexist – both are true. Reinscription – 'now I have written the above words' – triggers re-analysis.

This is an example of the truth-minefields which Gaskell lays, and it introduces at the start of the tale how language is to operate in breaking

92 Ibid., p. 220. A student first pointed out this 'triangularity' to me in a class discussion of this text – thanks to Martin Connor.
93 Ibid., pp. 220–1.

open and revealing buried scenarios. Gaskell's narrators are released into a form of linguistic self-betrayal or revelation, which often takes place through the slightest and slyest word. Paul continues 'But I have nothing to do with that now', yet – or as Gaskell might put it, *and* (for there is no necessary contradiction here) – he proceeds to tell a story full of these slippages and one which is entirely to do with the familial triangulations he claims to put behind him. It is not just that we are dealing with an unreliable narrator here; rather, as the selves of the novel uncoil and contradict themselves, the sacred notion of reliablility itself begins to unwind too (as is also true of Freud as narrator of the 'Dora' case). From an account of his father's well-meaning 'unconscious' double actions Paul navigates his own double-thought. Another's self-deception, or rather the points at which the incompatible strata of another self are laid bare, catalyses a double truth in the narrative's own voice and position. The way that one finds oneself speaking about another is one way in which one finds oneself returning to a disavowed account of the self. This is a form of narrative self-analysis at which Gaskell is adept, as she lets her voices run toward discoveries which come to be made entirely through the way that something is said, or not said.

But we are pursuing the hysteric through this tale, and Phillis is the focus here (although Paul is her flawed interpretor, as Freud is Dora's). If Paul feels himself cornered by his three-cornered room, Phillis is repeatedly viewed as the prime narrative object and the thing caught between the competing demands of the gazes of the characters – her father's, her suitor's, and that of Paul as her confidante. Phillis is the story's avid reader, yet from the start she is the thing to be 'read', looked at, interpreted. 'Keep your head still; I see a sketch,' says Holdsworth, a sketch, as it happens, of Phillis as Ceres ('he was holding some wheat ears above her passive head, looking at the effect with an artistic eye'). But more important here than how Holdsworth looks at Phillis is how Paul sees Phillis in a state of being looked-at:

> He began to draw, looking intently at Phillis; I could see this state of his discomposed her – her colour came and went, her breath quickened with the consciousness of his regard; at last, when he said, 'Please look at me for a minute or two, I want to get the eyes,' she looked up at him, quivered, and suddenly got up and left the room.[94]

Within the text, then, we have a narrator as reader (and sometimes misreader) of bodily signs. It is around this time that Phillis begins to display unusual 'symptoms' – breathlessness, her body refusing to do what she wants it to, her colour betraying what is really going on. At these

94 Ibid., p. 272.

moments, the verbally precise young woman begins to precipitate a linguistic failure which also affects and infects everyone else.

Phillis's growing awareness of herself as visual object thus takes in both her image as a desirable woman and her role as the subject of crisis desperately trying to fend off the visual interpetation of her body's symptomatological betrayals. Like Freud as reader of Dora in Marcus's essay, Paul is 'at once dogmatically certain and very uncertain',[95] unable to pin Phillis directly to a specific phase in development. The whole story, in fact, circulates this ambiguity between her childishness and womanliness, allowing its resolution to depend upon who is making the judgement at any one time. Our earliest glimpse of her is mediated by Paul, who says, 'I thought it odd that so old, so full-grown as she was, she should wear a pinafore over her gown',[96] a detail picked up on later when it is used to make a point about scale, Phillis's change of clothing constituting the major event of the autumn:

> I can only remember one small event, and that was one that I think I took more notice of than anyone else: Phillis left off wearing the pinafores that had always been so obnoxious to me; I do not know why they were banished, but on one of my visits I found them replaced by pretty linen aprons in the morning, and a black silk one in the afternoon. . . . this sounds like some book I once read, in which a migration from the blue bed to the brown was spoken of as a great family event.[97]

Such a minute event is the stuff of psychoanalysis, as is the almost imperceptible change of tone which comes with rereading a scene in terms of a subsequent desire. We have moved from Paul thinking the pinafores 'odd' to Paul finding them 'obnoxious', claiming that this had always been his position. There is no intervening signal of change, and so with his later approval of the change Paul is rewriting this history of his earlier opinion. Freud does this too in the 'Dora' case; as Marcus again writes,

> he is . . . utterly uncertain about where Dora is, or was, developmentally. At one moment in the passage he calls her a "girl," at another a "child" – but in point of fact he treats her throughout as if this fourteen-, sixteen-, and eighteen-year-old adolescent had the capacities for sexual response of a grown woman.[98]

This phenomenon is made explicit a little later in *Cousin Phillis*. Holdsworth is on the inside, indoors; Phillis goes outside and starts

95 Marcus, 'Freud and Dora', p. 78.
96 *Cousin Phillis*, p. 226.
97 Ibid., p. 247.
98 'Freud and Dora', ibid., p. 78

shelling peas. Paul follows her and they have a conversation. All the time, Holdsworth is watching, his gaze being itself unseen. When Paul returns to him indoors, Holdsworth declares Phillis a beautiful woman. 'Woman! beautiful woman!' continues Paul's narrative,

> I had thought of Phillis as a comely but awkward girl; and I could not banish the pinafore from my mind's eye when I tried to picture her to myself. Now I turned, as Mr Holdsworth had done, to look at her again out of the window: she had just finished her task, and was standing up, her back to us, holding the basin in it, high in the air out of Rover's reach.
>
> ... At length she grew tired of their mutual play, and ... she looked towards the window where we were standing, as if to reassure herself that no one had been disturbed by the noise, and seeing us, she coloured all over, and hurried away, with Rover still curving in sinuous lines about her as she walked.[99]

Again the scene is patterned around a triangle of desire with Phillis at its apex. There is a significant trading or borrowing of glances here, too: first Paul looks anew with the eyes of Holdsworth ('Now I turned, as Mr Holdsworth had done, to look at her again'), and then Phillis, hoping to reassure herself of her solitude, receives instead the look of the others and sees herself as the judged object for them which she must be through their eyes.

In a sense, the story charts Phillis's fall into the gap marked by everyone's uncertainty about her identity. The questions about her which the text raises (girl or woman?) are reinforced by those raised about her desire (father or stranger? – that old Freudian question, What does woman want?). As she slides into the impossibility enforced by these unanswered questions, she exchanges one form of speech for another, falling (away from Greek, Latin and the straight talk of the household) into inarticulacy and eventually silence, animated by an increasingly articulate body. It would be easy, then, to offer up this body and to produce from it a kind of Charcotian reading of its symptoms. What is more interesting is the way in which the text offers the body through its unresolved narrative. I have said that Phillis's 'fall' takes place after her sexual rejection, but this is not entirely true, for a more important factor in her crisis is the challenge to a straightforward relationship of language to meaning which Holdsworth (and to some extent Paul) brings with him from the outside world. The Holmans exist in an idyll which is not just rural, but also linguistic – they say what they mean and mean what they say, and there is ostensibly no gap between word and meaning. Paul comes from a world in which language behaves more deviously (he speaks at one point of 'arrang[ing his] manner of speech'[100]), and when Holdsworth speaks to Phillis in this 'style', she is confused, and silenced:

99 *Cousin Phillis*, pp. 261–2.
100 Ibid., p. 306.

> This was a style of half-joking talk that Phillis was not accustomed to. She looked for a moment as if she would have liked to defend herself from the playful charge of distrust made against her, but she ended by not saying a word.[101] . . . [Holdsworth's] tone of badinage (as the French call it) would have been palpable enough to any one accustomed to the world; but Phillis was not, and it distressed or rather bewildered her.[102]

Holdsworth's unreliable, playful words are both dangerous and seductive,[103] and it is words which are to precipitate Phillis's breakdown (words act against her in that they fester inside her, both in Paul's repetition of Holdsworth's love and in the written word about his marriage to another). Phillis has no mode of response, so she is silenced, and as the text proceeds her silence weighs increasingly heavily on everyone ('I felt obliged to say something,' says Paul later; 'it was stupid enough, but stupidity was better than silence just then'[104]). She begins to slide away into animalistic gestures which Paul is forced to read and respond to ('her white looks, her weary eyes, her wasted figure, her struggling sobs . . . She was making a low moan, like an animal in pain, or perhaps more like the sobbing of the wind'[105]). The scene in which Paul speaks of Holdsworth's love is an agony of suppressed expression, with Phillis failing to say more than 'Don't – ', Paul inserting words into the space of her silence, and desperately trying to interpret the spectacle of gestures which her body is becoming – blushing, choking, crying, 'her tender mouth . . . curved with rapture', and finally fearing that 'her face expressed too much, more than the thankfulness to me she was essaying to speak'.[106] As a chasm is gouged open between word and meaning, word and thing, Phillis can only proceed to speak 'honestly' with the body.

Cousin Phillis, as I have said, is in many ways a small text, looking with a scaled-down eye at the telling minutiae of family life. Brimful of tiny gestures read, acted upon, or missed in the interpersonal exchanges of the drama itself, it also spills over with physical signs, gestures, involuntary expressions which Jacques Lacan calls 'a *remainder* that no analyst will neglect, trained as he is to retain whatever is significant, without always knowing what to do with it'.[107] Roland Barthes, after Heidegger, writes that 'language speaks', to indicate the possessing, involuntary

101 Ibid., p. 259.
102 Ibid., p. 270.
103 Even the Minister listens: 'there is a want of seriousness in his talk at times, but, at the same time, it is wonderful to listen to him! . . . I listen to him till I forget my duties, and am carried off my feet' (ibid., p. 266).
104 Ibid., p. 272.
105 Ibid., p. 283.
106 Ibid., p. 285.
107 'Seminar on "The Purloined Letter"', in *The Purloined Poe: Lacan, Derrida, and Psychoanalytic Reading*, ed. John P. Muller and William J. Richardson (Baltimore, MD, and London: Johns Hopkins University Press, 1988), p. 30; Lacan's italics.

qualities of a language which 'means' quite apart from the would-be conscious or moral purposes of authorial control. In *Cousin Phillis* the language which speaks what the subject cannot directly articulate through her mouth in sentences of sense also comes from 'elsewhere': bodies speak when words fail. In Phillis's mute agony is an abundance of signfication. When the news of the marriage becomes public, Phillis's body again speaks:

> I could not help looking just for one instant at Phillis. It seemed to me as if she had been keeping watch over my face and ways. Her face was brilliantly flushed; her eyes were dry and glittering; but she did not speak; her lips were set together almost as if she was pinching them tight to prevent words or sounds coming out.[108]

Then follows an awful family dinner-table scene, in which Phillis can only use words as agents of deferral, not to speak the truth but precisely to prevent it being spoken, keeping the discussion 'off the one raw surface, on which to touch was agony'.[109] It is here that the real nature of the family's breakdown becomes clear. Before lapsing into silence altogether, Phillis engages in one final linguistically desperate act, using words to say *not* what she means, but quite the opposite. Language then becomes not an agent of communication, but a means of self-defence, a screen erected to prevent self-revelation. There are many references in the text to the loss of an old language and the coming of a new,[110] but what is

108 *Cousin Phillis*, p. 301.

109 Ibid., p. 302.

110 A powerful parallel is also being played out here with the model of the clash of languages suggested by Sándor Ferenczi, one of Freud's closest friends and his pupil. His 1932 paper, 'Confusion of Tongues between between Adults and the Child: The Language of Tenderness and the Language of [Sexual] Passion,' dramatizes a mismeeting of adult desire and infant love which, for Ferenczi, accounts for 'seduction'. Because the two sides speak on different terms, what each communicates to the other is read awry: 'An adult and a child love each other; the child has the playful fantasy that he will assume the role of the mother to the adult. This game may also take on erotic forms, but always remains on the level of tenderness. This is not true of adults with a pathological predisposition ... They confuse the playfulness of the child with the wishes of a sexually mature person . . .' (collected in Masson's *The Assault on Truth*, pp. 291–303, quotation p. 297). Through a more expanded discussion of this theory we might identify the Holmans' original language as the language of tenderness, and its breakdown as entry into the language of passion. This is taken up by Jean Laplanche, whose more recent work on deferred action will be the main focus of Chapter 4. Laplanche has worked through the history of psychoanalytic theories of seduction, and negotiates an alternative path to that traced by Jacques Lacan on the child's entry into language, by taking Ferenczi as its starting-point. In particular, Laplanche's model of the 'enigmatic signifier' suggests that there can be a gap between word and meaning which is not simply a shift, which underpins the confused relationship between adult and child. See in particular Chapter 3, 'Foundations: Towards a General Theory of Seduction', in *New Foundations for Psychoanalysis* (Oxford: Basil Blackwell, 1989).

important is the way in which this linguistic change is entirely focused on Phillis' crumbling psyche. 'Until now', Paul writes,

> everything which I had heard spoken in that happy household were simple words of true meaning. If we had ought to say, we said it; if anyone preferred silence, nay if all did so, there would have been no spasmodic, forced efforts to talk for the sake of talking, or to keep off intrusive thoughts or suspicions.[111]

As 'simple words of true meaning' are lost, so Phillis begins to speak 'otherwise', and the narrative points our gaze towards her hands. After her first outburst against her parents, she reconciles herself with a 'tender pantomime' of hand-holding with her mother. After she slips into the new register which enables the deception above, hands speak more 'honestly' than words:

> But once my eyes fell upon her hands, concealed under the table, and I could see the passionate, convulsive manner in which she laced and interlaced her fingers perpetually, wringing them together from time to time, wringing till the compressed flesh became perfectly white. What could I do? I talked with her, as I saw she wished; her grey eyes had dark circles round them, and a strange kind of dark light in them; her cheeks were flushed, but her lips were white and wan. I wondered that others did not read these signs as clearly as I did.[112]

Is Paul then here engaged in some form of pre-theorized psychoanalytic 'reading' of Phillis (taking us with him), and if so, is psychoanalytic criticism itself then merely about 'reading signs' – the signs of aberrational behaviour, or of linguistic slippage which betrays a telling psychic discord – signs even of silence? Feminist film critic Tania Modleski quotes Hélène Cixous near the beginning of her analysis of 'Time and Desire in the Woman's Film'. Modleski focuses on Cixous's work on utterance by quoting Cixous's 'Silence is the mark of hysteria': 'The great hysterics have lost speech, they are aphonic, and at times they have lost more than speech. They are pushed to the point of choking, and nothing gets through.'[113]

Modleski comments:

> It seems fair to say that many of the classic film melodramas from the 30s through the 50s are peopled by great, or near-great hysterics – women possessed by an overwhelming desire to express themselves, to make themselves

111 *Cousin Phillis*, 302.

112 Ibid., p. 304.

113 Hélène Cixous, quoted by Tania Modleski, 'Time and Desire in the Woman's Film', in *Film Theory and Criticism*, ed. Gerald Mast, Marshall Cohen and Leo Braudy (Oxford: Oxford University Press, 1992), p. 537.

known, but continually confronting the difficulty, if not the impossibility of realizing this desire.[114]

That hysteria is first addressed by the cathartic 'talking cure' and then by Freudian free association and dream analysis underlines its role as a condition occupying an uneasy space between speech and silence, a space which is curiously cultural, transgressing the boundary between the 'purely' psychological and that which concerns language, spectacle, representation. If the hysteric's need to speak is countered by the impossibility of utterance, and the woman is caught between the masculine identification of speech and the feminine identification of silence, what form of expression is left? How, then, do we read the expressions which lie between clear speech and absolute silence?

In the next chapter I want to develop some of these questions, but through a more specific sketch of how developments in psychoanalytic theory have infiltrated critical practice, and vice versa. Moving away from the figure of the hysteric for the present (although she will continue to return in a number of guises throughout this book), I will use the strange history of readings of Edgar Allan Poe as my next frame of analysis, turning specifically to readings which have orbited the work of Jacques Lacan.

114 Ibid.

2

Psychoanalytic Criticism and the Case of Poe

The work of Edgar Allan Poe has served as a touchstone for shifts in psychoanalytic criticism throughout this century. To the bemused reader, encountering the vast array of critical interpretations which have developed around his work, it might seem that Poe had become all things to all critics. In particular, the critical body which clusters around the story 'The Purloined Letter' forms a dossier of debates within psychoanalytic literary interpretation. Slice into Poe criticism and you get a cross-section of how discussions have crystallized into the strata of psychoanalytic history. In this chapter I want to present a range of the possible and most prominent ways of psychoanalytic reading with reference to one distinct focus – the works of Poe. Some of the approaches to Poe discussed here are certainly more familiar than others – Marie Bonaparte's psychobiography, for instance, or the debate around the 'Purloined Letter' story between Jacques Lacan and Jacques Derrida, mediated by Barbara Johnson. Less familiar routes into his work might be taken through the death drive and the dead woman, or through the intertextual focus of Roger Corman's cycle of Poe films in the early 1960s, which open up psychoanalytic significances solely through the imagistic possibilities of cinema.

That Poe should have become a special object of psychoanalytic interest is not surprising. When his stories and poems are not concerned with mental aberration, decay or the indefinitions of sexual identity, they focus on the exquisite impossibility of fixed knowledge in a world in which truth is an endlessly circulating, maddeningly unfixed ideal or image, never to be pinned down or guaranteed. Even death is not certain. Ligeia (in the story of that name) dies but doesn't die, merging into the ghost of Rowena (the woman who succeeds her as the narrator's love after Ligeia's death). The boundaries between women, even between opposite female identities, begin to break down as an effect of the narrator's fantasy: Ligeia the black-eyed, raven-haired beauty metamorphoses into Rowena, her fair-haired and blue-eyed successor. Rowena dies to become the already dead Ligeia, and Poe finds 'the most poetical topic in the world', 'the death . . . of a beautiful woman', not once but twice – an issue which recently Elisabeth Bronfen has discussed in terms of the figuration of woman in the death drive.[1] The dead Ligeia returns in the corpse of the dead Rowena; in their undifferentiated death, one supplants and becomes the other, yet at the same time as death is transgressed and poses no limit, still both women are subject finally to the limits and desires of their male viewer and narrator. Conjured into a precarious post-mortem existence built only on *his* mourning and fantasy, even the last affirmative line of the tale is stricken with the *question* of Ligeia:

> 'Here then, at least,' I shrieked aloud, 'can I never – can I never be mistaken – these are the full, and the black, and the wild eyes – of my lost love – of the lady – of the LADY LIGEIA.'[2]

Poe is adept at turning sure statements into shaky questions, endlessly circumnavigating conviction throughout his work, so that the effect of this conclusion is only inconclusiveness: *can* he never be mistaken?

This is only one example of how Poe's heroes are never allowed to know their experience with the assurance of realist ideals – he takes the gothic model of psychic uncertainty and builds it into a brooding, psychosexual principle of self-division. His characters seldom lie, but their inability to utter straightforward, mutually shared truths is even more precarious than falsehood, exposing the slippery foundation of utterance. Nothing is guaranteed, and there are few rules; those who survive this overwrought, delicate immoralism are pedantic, absurd, decadent. Roderick Usher (in 'The Fall of the House of Usher') is perhaps the most famous example of this, the text's pathologically sensitive (male) hysteric, whose 'morbid acuteness of the senses' is presented in exquisite detail, though its cause is not pinpointed except as a vague family malaise ('a constitutional and a family evil'[3]). The family gone wrong is also at stake in his sister Madeline's condition, and Poe lists the symptoms as clearly as he evades the causes: 'A settled apathy, a gradual

1 *Over Her Dead Body: Death, Femininity and the Aesthetic* (Manchester: Manchester University Press, 1992). Although I will refer briefly to Bronfen in Chapter 5 on the death drive, her reading of Poe (scattered across that text) offers a significant recent intervention in debates around his work.
2 'Ligeia', in *Selected Writings* (Harmondsworth: Penguin, 1984), p. 126.
3 'The Fall of the House of Usher', in ibid., p. 143.

wasting away of the person, and frequent although transient affections of a partially cataleptical character, were the usual diagnosis.'

The house too is infected, and in that it stands as the real body of the family (the '*House* of Usher' is both the family line and the building itself), it also infects: 'An excited and highly distempered ideality threw a sulpherous lustre over all'.[4] Poe pushes psychological conventions with which we might be comfortable until they capitulate, until we are supremely discomforted; he turns images of the subject inside out so that the diseased subject becomes the image. The mad woman may lurk in the attic, in the cellar of the self; the open window might betray the flimsy division of inside and out; thresholds are transgressible. But it is the self which is an image for the house rather than the other way around; it is the Ushers who are subject to and infected by the family seat, as the first (Roderick and Madeline) become the cracked image and metaphor for the second (the house itself). And if it is all a nightmare, a dark fantasy, and to be dismissed as such, Poe's narrator informs us in the first paragraph that it is *this* which is most real. The most awful possible fantasy is thus posited as an image of dismal reality, the revelation behind the veil, the clear image revealed after the narcotic haze has lifted:

> I looked upon the scene before me . . . with an utter depression of soul which I can compare to no earthly sensation more properly than to the after-dream of the reveller upon opium, – the bitter lapse into everyday life – the hideous dropping of the veil.[5]

Do not be comforted that you are about to read a story reassuringly and monstrously *other*, this reminds us: it is primarily *here*, in the macabre vision set before the narrator, in the bizarre stuff of the story itself, that banal reality is situated (in 'the bitter lapse into everyday life'), beyond the 'goading of the imagination'.[6] And yet, with this warning, Poe launches into a story obsessed with artifice, not an opium-fantasy but a fabricated construction. It is high camp in its ritualistic decadence, self-conscious in its almost morbid awareness of its own fictiveness.

It is this which proved most attractive for cult horror-film director Roger Corman in his cycle of Poe adaptations made in the early 1960s, infused with a psychedelic palette and presented through cinematography which stressed the pathology not just of characters but of the material worlds they inhabit. Extreme psychic sickness is foregrounded throughout the films, and the inextricable connection between unnatural familial relationships and pathology is made in each film. Corman reads

4 Both quotations, ibid., p. 145.
5 Ibid., p. 138.
6 Ibid.

Poe for a Freudian resonance which has none of the crudity of those written ('vulgar Freudian') critical analyses which stress the same Oedipal elements in text-book form (we will look at a few later in this chapter). By the mid-twentieth century it had become increasingly hard to interpret Poe's stories outside of a Freudian agenda which the stories themselves pre-date. What then is a 'Freudian reading', and what governs the form it can take as well as the arguments it can make? In answering this by offering Corman alongside Marie Bonaparte or Jaques Lacan, I am assuming that a film can be a 'reading' too. In his highly psychologically inflected interpretations of these stories, Corman produces a set of visually inventive, filmic psychoanalytic readings which I briefly want to set alongside the more conventional forms of 'reading' with which this book is generally concerned.

The House of Usher (1960) foregrounds its own artifice in a stark but limited technicolour palette of blues, reds and whites which only looks right when the central dream-sequence, into which the film's hero slips, takes off: only dreams, cinematically speaking, look like this, only in dreams does the lurid 'way of seeing' which infects the whole film, make sense. Corman elaborates the source text in a number of ways which open up its unspoken implications. Where the original story establishes the narrator as Roderick's childhood friend, with Madeline the dark secret of the household, the film posits her as member of a lively Boston social scene and contemporary, as well as fiancée, of the narrator. Vincent Price, meanwhile, plays Roderick more as an aged father-figure and sexual rival in the contest for Madeline's love, bringing Poe's less overtly played point about family incest centre-stage. Roderick the brother becomes also the father, and Madeline's acceptable connections with the real social world only serves to *normalize* her family romance. In the final reel, Roderick drops his pistol in a poised ecstasy of masculine submission as his phallic (and by now dead) sister advances toward him, turning murder into incestuous consummation.

Corman's Poe knows that repression is about something being buried alive, and *The Pit and the Pendulum* (1961) is particularly interesting from this point of view. It takes the basic premise of the original story of 'The Pit and the Pendulum' – itself simply a sequence of torture-rituals, in which the narrator talks us through his agony at the hands of the Spanish Inquisition – and works it together with a repeated reading of the incest motif of 'The Fall of the House of Usher'. The problems of authorial centring, of Poe as final guarantor of the unconscious meaning of the text, which have haunted 'psychobiographical' readings, are never so clear-cut as in the analysis of Poe which takes place in these films. Whilst Hollywood is adept at taking 'original' tales and turning them into unrecognizable (and rather sweeter) films, what Corman does, in presenting a collage of psychosexual positions culled from a range of Poe texts, is more interesting. As reader of Poe, Corman directs a cinematic interpretation which may emphasize unconscious elements but does not pin these down to, or privilege, a single fixed point of view or authorial source. The films reread the stories in a manner which emphasizes their curious morbidity, but this cannot be finally offered up as a work of psychobiography – Poe's unconscious desires lying 'beneath' the tales are *not*, then, at issue, although the unconscious *resonances* of the tales *are*. Weaving together elements from two texts, *The Pit and the Pendulum* presents another drama of active repression and twisted Oedipality. Haunted by the possibility that he buried his wife alive, Nicholas Medina (Vincent Price) goes slowly mad in the manner of Roderick Usher (whom Price had played in the previous film), fainting, weeping, hallucinating, and exuding 'an air of definite guilt', as his sister says early in the film. '*We have put her living in the tomb!*', Usher cries at the close of Poe's story, just as she is returning from the dead, the avenging angel who incestuously takes her brother/lover to his death. Similarly, 'Nicholas believes that Elizabeth may have been interred prematurely . . . because of what happened to his mother' says the doctor of Corman's film: 'She was walled up in her tomb whilst yet alive.' And just like Poe's Usher family, Corman's Medina family are infected with the poison of incest, when the generations collapse into and relive the experience of each other. The Medina Castle, like the Usher House in story (and the previous film), is the family's real body, its inhabitants just the individual symptoms of a collective disorder.

There is, then, no escape from the perverse family here, and all relationships take place within its sexual domain: Nicholas is the son of a Spanish Inquisition torturer and, with his sister (Catherine, played by Luana Anders), continues to inhabit the bizarre world of the father which is his inheritance – the castle, the torture chamber, his psychic state, by turns frenzied and delicate: 'This was my father's world. Am I not the spawn of his depraved blood?' The 'blood' infects his wife Elizabeth (Barbara Steele) who, by her proximity to Nicholas and the chamber, turns into his mother, becoming guilty of a family history which her existence repeats. Indeed, in true Poe fashion Corman ensures that visual connections between characters enforce the possibility that everyone in this film is related to (and desires) everyone else. The film begins with Francis (John Kerr) seeking his sister Elizabeth, and finding her entombed with Nicholas and *his* sister. Mothers fornicate with uncles, and though direct incest is not actually committed through any 'wrong' connection between Nicholas, Catherine, Elizabeth and Francis, it is implied by the siblings' interrelatedness through marriage.

If Poe stresses the sexual interconnection of generations, Corman makes it flesh through casting. In flashback, Nicholas tells of witnessing his father's murder of his mother, in which he (Price) plays the role of the father dispatching the mother. If Poe *suggests* the identification of sons with fathers, Corman realises it. In flashback, we see Nicholas as a child entering the chamber, in a story first told by Catherine:

> He had been forbidden by our father to enter this chamber at any time. The curiosity of youth, however, overcame his fear at challenging our father's discipline. He *had* to see the chamber.

He sees more than this: in flashback, his father (also played by Price) kills his mother in a sexual frenzy of torture – the adult Nicholas's eyes replay the image of terrible desire which is that of himself-as-father entombing the mother in a jealous desire which kills her. So Nicholas kills his wife (not once but twice), the wife who has taken the role of the mother – in becoming his own father, he is able to collapse wife into mother and kill them both. Thus Corman delivers a twisted rendering of the Oedipal myth, which identifies son with father and torture with sex – killing the mother-surrogate is as near to the father's sexual relationship with the mother as the son can get. Roderick Usher only loves his sister; in turning Roderick into Nicholas, Corman makes the familial desire of sibling incest into a more classically Freudian desire which crosses generations. If Freud's Oedipus is tormented by anxiety and guilt, Corman's shifts the terms of the original myth and makes parricide matricide, murder being the nearest the horror protagonist can get to sex. Killing the father to love the mother thus becomes killing the mother *as a way of* loving her. The son supplants the father, and murder becomes a displaced desire – matricide, then, *is* incest.

Corman's reading draws out from Poe the terrors implicit not just in extreme gothic circumstances but in the family structure itself. Abnormal as Poe's cast of characters might be, in giving them this particular cinematic life Corman, for all his visual extremity, manages finally to stress not utter otherness but rather their proximity to the 'normal' taboo passions of the Freudian world. These are issues at stake in a number of textual readings of Poe, to which I now want to turn. Poe's texts endlessly negotiate in style and narrative the interconnection of fiction and madness, but they also, as we shall see, foreground and facilitate psychoanalytic readings coming from different directions, as spaces within which the implication of psychoanalysis in the literary construction of the self is played out.

In her account of 'The Poe-etic Effect',[7] Shoshana Felman has chronicled criticism of Poe, tracing a literary case history in the volatile,

[7] 'On Reading Poetry: Reflections on the Limits and Possibilities of Psychoanalytical Approaches', collected in Felman's 1980 volume, *The Literary Freud: Mechanisms of Defense and the Poetic Will*; also collected in *The Purloined Poe: Lacan, Derrida and Psychoanalytic Reading*, ed. John P. Muller and William J. Richardson (Baltimore, MD, and London: Johns Hopkins University Press, 1988), pp. 133–56. Since interested readers might find it easier to trace as many references as possible to this one useful volume, page numbers refer to this text. All italics in subsequent quotations are Felman's own from the original text.

self-betraying scholarship which Poe's corpus has engendered, from Joseph Wood Krutch's 1926 image of Poe as exemplary neurotic (*Edgar Allan Poe: A Study in Genius*)[8] to Marie Bonaparte's seminal study of 1933, *The Life and Works of Edgar Allan Poe* and beyond. According to Felman, Krutch's Poe is 'a pathological condition of sexual impotence',[9] and Poe's writing is reduced 'to an abnormal condition of the nerves'[10] – a 'study in genius' is thus more of a study in authorial madness buried in the text. Poe is sick, the text is thick with symptoms, and it is up to the dispassionate (healthy) reader to make the correct diagnosis. For Freud, however, there is no ideal model of an achieved and healthy sexual identity at which we are all capable of arriving and sustaining through harmonious adulthood. Indeed, one of the strongest arguments which feminism has taken from psychoanalysis is precisely this: 'failure' of identity, as Jacqueline Rose puts it, 'is not a moment to be regretted in a process of adaptation, or development into normality, which ideally takes its course':[11]

> The unconscious constantly reveals the 'failure' of identity. Because there is no continuity of psychic life, so there is no stability of sexual identity, no position for women (or for men) which is ever simply achieved. Nor does psychoanalysis see such 'failure' as a special-case inability or an individual deviancy from the norm. . . . 'failure' is something endlessly repeated and relived moment by moment throughout our individual histories. It appears not only in the symptom, but also in dreams, in slips of the tongue and in forms of sexual pleasure which are pushed to the sidelines of the norm.[12]

In the light of this, Krutch's account of ideal, achieved psychic health seriously distinguishes his work from the body of psychoanalysis, however strong his claims to allegiance. Krutch's Poe is aberrational precisely because he demonstrates the 'failure' of identity so flagrantly. Thus Poe's texts do not simply *contain* or represent monsters; instead, his whole corpus is an exercise in psychic monstrosity from which we 'normal' ones can only learn. The key difference operating in Krutch is then marked out between text and reader: this is literature as a psychic-moral fable which can only assure the reader of her normality. For Felman, as we shall see towards the close of this chapter, these responses opened up a history of readings significant for their extremities and disagreements, which is itself indicative of how the text operates on and provokes its readers.

8 Joseph Wood Krutch, *Edgar Allan Poe: A Study in Genius* (New York: Knopf, 1926).
9 'On Reading Poetry', p. 138.
10 Krutch, quoted by Felman, ibid., p. 139.
11 *Sexuality in the Field of Vision* (London: Verso, 1986), p. 91.
12 Ibid., pp. 90–1.

I

Bonaparte on Poe: Psychobiography and Dream Analysis

One of the most important early examples of psychobiographical reading is Marie Bonaparte's full-length study, *Edgar Poe: Étude psychoanalytique* (Paris 1933, translated into English in 1949 as *The Life and Works of Edgar Allan Poe: A Psycho-analytical Interpretation*), a pioneering analysis of the relationship between individual pathology and textual symbolism, and one of the more substantial examples of the critical 'school' which came to be known as 'pathography'. Once more, the aim and effect of this approach is diagnosis, but Bonaparte's emphasis is less judgemental of 'literary sickness'. In this reading, Poe's texts are encrusted with a personal symbolism which can be decoded to offer a primary image of the author's mental state, which itself has held a hypnotic power over readers because there, but for the grace of successful repression, go they. For Bonaparte, Poe's pathology is our own writ large; his stories are the morbid magnification of a 'condition' which, Freud stressed, is still intrinsic to everyday life. It is, then, appropriate that Bonaparte chose for her basic critical tool the methods and models of Freudian dream analysis, developed to account for a *common* event (dreams themselves) in the mental life of everyone which, like parapraxes, testifies to the fact of repression and the existence of the unconscious (as well as to the 'failure' of seamless, controlled identity of which Rose writes above). Whereas Krutch's Poe is important because of the lessons in pathology his derangement offers to his 'healthy' readers, Bonaparte's Poe intensifies the effects of the traumas at the root of 'normal' psychic life until they explode into raging literary symptoms – the morbidly neurotic self, unfettered and demonstrating its unthinkable desires between the lines of the page.

In essence, Bonaparte offers what Felman calls 'a clinical "portrait of the artist"'.[13] Under this analysis, Poe's stories and poems emerge primarily as a symptomatizing body, offering traces of a real human crisis which lies behind it in the psyche of the author. Thus the codes which Bonaparte finds in the text are plotted as a devious pathway through to the aberrational mentality which lies behind it. For Bonaparte, Poe works on his texts by working on himself through his characters, identifying with them, splitting himself across them, and embodying himself in them. It is the psychoanalytic reader's role to

13 Felman, 'On Reading Poetry', p. 141.

> determine the extent to which the author's personality, split into psychic elements seeking to embody themselves in different characters, permits the author to re-embody himself in each of the characters observed.[14]

The familar Oedipal models are there in the characters, as life is mapped onto text and text in turn becomes an alternatively written unconscious life, with various characters and even material things (buildings, landscapes, objects) standing in for the mourned consumptive mother Elizabeth Arnold, the long-dead father, or the ailing, unthreatening child-bride of thirteen (Virginia) whom Poe married as a convenient means of keeping faith with the mother. So the topography of morbid pathology and necrophilic obsession is mapped out through a text read as dream, and characters' ambiguities are 'solved' into psychic roles or imagos.

This form of author-centred psychobiographical reading has been roundly criticized in debates since the 1960s, but I do not want simply to reproduce this critique. Instead, I shall briefly outline some important points in Bonaparte's analysis, not as an example of the maligned form of psychobiography which critics would argue is now only of interest to historians of abandoned critical theories, but as a way into Freud's technique of dream analysis (which Bonaparte takes as her model), as well as a moment in the debate around 'applied psychoanalysis'. Bonaparte's work on Poe follows the methods of Freudian dream analysis fairly closely, as have other psychoanalytic readers. 'Interpreting means finding a hidden sense in something,' writes Freud in the *Introductory Lectures*;[15] and so for Bonaparte the text forms the manifest dream, and it is the work of the critic to uncover its latent content, with reference to external biographical sources (information about the death of the parents, say, or of Poe's own marriage). To do this Bonaparte must reach an understanding of how the text operates as the product of a form of 'dream-work', which to Freud is the mental process which transforms latent dream-materials into the manifest dream. Just as the dream-work connects these two (latent and manifest) levels of the dream, causing the analyst to read through the manifest image *back* to the latent meaning, so the literary analyst must, for Bonaparte, recognize that the written text is also always double:

> [I]t must not be forgotten that though, on the surface, a literary work relates a manifestly coherent story, intertwined with it and simultaneously, another secret story is being told which, in fact, is the basic theme. Though, therefore, the manifest tale normally obeys the rules of logic, this deeper current is subject to other laws.[16]

14 Marie Bonaparte, *The Life and Works of Edgar Allan Poe: A Psycho-Analytic Interpretation* (London: Imago, 1949), trans. John Rodker, p. 639.
15 p. 115.
16 *Poe*, p. 654.

This is what Freud calls 'seek[ing] to find the essence of dreams in their latent content and in doing so . . . overlook[ing] the distinction between the latent dream-thoughts and the dream-work'.[17] In her discussion of Bonaparte in *Psychoanalytic Criticism*, Elizabeth Wright cites this quotation from Freud, pointing out: 'The problem [with Bonaparte] is not so much in what she does (which is very interesting), but in what she claims: that this is where the true meaning is to be found: '[It] is as though Poe himself were to declare "*Because* I am still fixated on my mother, I cannot love another woman"'.[18]

How is this argued in Bonaparte's text itself? Writing of the motif of the 'live-in-death mother', she traces Poe's displacement of affect attached to his mother onto certain characters:

> Berenice, Morella, Ligeia, Madeline, are as morbid, as evanescent as advanced consumptives, while their sylphlike motions seem, already, to exhale an odor of decay. Nevertheless, this simple displacement served to keep Poe ignorant, as for almost a century his readers, that these ailing sylphs were but forms of Elizabeth Arnold.[19]

The 'ignorance' of the author (and his non-psychoanalytic readers) of his true motivation is the veil which Bonaparte will tear down, revealing the truth of the text. The dream-work then applies itself to basic material, which it has to convert into something else before the dream can enter the pre-conscious (the realm in which we experience dreams) – 'the dream as a whole', writes Freud in the *Introductory Lectures*, 'is a distorted substitute for something else, something unconscious, and . . . the task of interpreting a dream is to discover this unconscious material'.[20] This is Bonaparte's task too, and the figures in Poe's subjective landscape do not walk undisguised through the pages of his tales, but take on the cloak which psychic processes lend them. Here, displacement of one figure into the image of another is at work, and Bonaparte goes on to suggest that even Poe's buildings are cast in part as displaced forms of the mourned mother – it is not just Madeline who in decay takes her role, but the whole House of Usher itself, 'a house whose walls, whose atmosphere, breathe putrefaction. To effect this gross displacement, Poe employs one of man's universal symbols – that which represents woman as a building.'[21]

For Bonaparte, Poe's mother-obsession is then an incestuous necrophilia. The Oedipal mother figures centrally in psychobiographical

17 Freud, *SE*, vol. v, p. 506 n., also p. 580.

18 Elizabeth Wright, quoting Bonaparte in *Psychoanalytic Criticism: Theory in Practice* (London and New York: Methuen, 1984), p. 41; her italics, Bonaparte, *Poe*, p. 655.

19 Bonaparte, *Poe*, p. 642–3.

20 p. 144.

21 *Poe*, p. 643.

readings, but it is not just the mother who is the object of Poe's taboo affections in story after story: it is the *dead* mother. Bonaparte stresses (and who could not) the massive morbidity of Poe's corpus – its corporeal obsession with corpses, with the processes of mental and physical decomposition, but particularly in the bodies of women. Emaciated, vengeful, their state of decay ever an index of their desirability, Poe's women are nevertheless finally and resolutely *dead*, even if reanimated. On these terms Bonaparte's can be no ordinary psychobiographical reading, such as one might construct around the figure of D. H. Lawrence through the living Oedipality of Paul Morel's relationship with his mother in *Sons and Lovers*. The death drive is laid onto Oedipus, through an account of the stories as acts of morbid compulsive repetition, as tale after tale compulsively and unconsciously revisits the moment of the mother's demise in Poe's own infancy. These are then also narratives of male masochism, as the repetition is carried out with guilt, desire and ghastly relish, an act of literary mourning charged with desire and never to be completed.[22] Mothers are also displaced in more peculiar forms here, as horses, ships, the earth and the ocean itself, as well as in architectural features to which Poe gives specific attention (such as chimneys), whilst the cell in which the pendulum swings over the pit has a barely disguised significance:

> Innumerable are the displacements which went to construct *The Pit and the Pendulum* nightmare. The cell as the contractile womb of the mother, the vaginal pit, and the penis-sythe of Time, are but the most striking.[23]

Bonaparte is then doing to the stories what Freud had done to dreams – these are Poe's nightmares writ large, and we can read them as Freud had read Schreber's text, effecting a living interpretation of an elaborate symbolism constructed through various processes of disguise or transformation. Analysis then reverses the dream-work:

> [T]he work which transforms the latent dream into the manifest one is called the *dream-work*. The work which proceeds in the contrary direction, which endeavours to arrive at the latent dream from the manifest one, is our work of interpretation. This *work of interpretation* seeks to undo the dream-work.[24]

Chapter 6 of *The Interpretation of Dreams* – a large and central section of the book – is dedicated to close discussion of the dream-work, and here Freud identifies its four principal mechanisms, through which latent thoughts are made into the dream-scenarios in the unconscious: *conden-*

22 This is also partly Elisabeth Bronfen's concern, not as a psychobiographer, but as analyst of the bodies of women which litter cultural history, and specifically Poe's writing.
23 Bonaparte, *Poe*, p. 645.
24 Freud, *Introductory Lectures*, p. 204; his italics.

sation, displacement, considerations of (or regard for) *representability* (through which the dream-work transforms latent thoughts into visual images), and considerations of *intelligibility* (or secondary revision – the process through which a more or less coherent scenario is constructed, which eliminates some incoherent or absurd elements). In Bonaparte, these processes take on a literary life (we have seen some examples of how she identifies displacement at work), and armed with the right tools Poe's latent desires can be detected at play in the manifest content of the text-as-dream:

> Who ever would have found his way through all this but for the keys, the laws, revealed by Freud in his *The Interpretation of Dreams?* . . . Indeed, all representations by courtiers, princes, or kings, of the parents we knew as children, as in "Hop-Frog" or "The Red Death," are so many displacements designed to render them unrecognizable for what they are, so that, unsuspecting, they may play their "guilty," libidinal parts.[25]

Despite Freud's assertion that the transformations of the dream-work upon the raw material of the dream's latent content is not strictly a creative process, this is the work which Bonaparte then takes as her model for *creative* life. Just as latent becomes manifest in the construction of a dream, so the unconscious of a literary work 'is able to pass into the conscious product of the written work'[26] through the agency of creativity, acting on the text's 'raw material' just as the dream-work does. This not only presupposes that a text has an unconscious, the writer's own, on which he or she can 'work', but that the process of translating this into literature is as accessible to the psychoanalysing critic as dreams are to the analyst. We have seen how analysis seeks to overcome the resistances and censoring in which the speaking analysand engages during free association. Bonaparte treats the written text as a similarly codified system which marks the difference between conscious and unconscious life, but through which the unconscious can also be read – as long as we know how to crack the code. The dream (and the text) is not, then, the unconscious laid bare, but an image of unconscious desire presented in a codified form – this is why Freud calls dreams 'The Royal Road to the Unconscious' – they offer a *way of approaching* latent desires which cannot be directly accessed. Via the 'surface movement and gleams of happenings' which find their way to the surface of Poe's text, Bonaparte thus gains a devious form of access to the 'real meanings' below:

> '[J]ust as our capacity for external perception, via the senses, can only perceive phenomena without probing their essence, so our faculty of inner perception

25 *Poe*, pp. 643–5.
26 Ibid., p. 641.

> can only observe surface movement and gleams of happenings in the inaccessible depths of our unconscious. Thus our conscious ego is never but the more or less watchful spectator of ourselves.[27]

In addition to the *displacement* of biographical figures into literary phenomena, Bonaparte also accounts for other images in Poe through the dream-work processes of *condensation, intelligibility* and *representability.* The consideration of intelligibility, or the process of 'secondary revision' of the latent elements worked on by the primary processes of displacement and condensation, ensures that dreams have a basic logic or narrative, as disparate elements are structured into a comprehensible *form.* Through this aspect of the dream-work, the mind eliminates the incongruous morass the latent dream has become, filling in gaps and creating a more or less coherent (if still sometimes absurd) sequence – a dreamed story we remember, connected to other dreams before and after in some sort of intelligible flow. Because literary texts have for Bonaparte (who does not read 'difficult' or avant-garde writings) to be intelligible or to offer a basic logic to justify events, they already impose some kind of order on chaos (texts are 'incessantly at work criticising and constructing in order to fit, to our most deeply repressed desires, that conscious, logical, and aesthetic façade which we call creative writing'[28]). Bonaparte does, however, show that other dream-forms of causality and sequence are also at work beneath the manifest surface of narrative. In condensation, latent elements are as Freud writes 'combined and fused into a single entity in the manifest dream',[29] and Bonaparte takes this up through Poe's creation of what she calls 'composite figures', which stand for and do the work of a number of figures in psychohistory. This discussion perhaps accounts for Bonaparte's most inventive readings, and characters are not simply composite but may also be split, with one 'real' person (including the authorial ego itself) becoming disseminated across several characters or a number of subject positions in the text:

> In *Morella, Ligeia,* and *Eleonora,* the manifest forms of the first wives begin as condensations of the images of Elizabeth and Virginia; they then, however, split to represent, separately, once more distinct from each other, the two images originally separate in the latent thought of the tale.[30]

Finally, the dream-work's consideration of representability, its reworking of fluid thoughts into concrete, visual images, takes place in Bonaparte's reading when Poe personifies or dramatizes desire through a 'visual' process. 'A Descent into the Maelström' (of which more later),

27 Ibid., p. 642.
28 Ibid., p. 663.
29 *Introductory Lectures,* p. 205.
30 *Poe,* p. 650.

which is a simple, apocalyptic tale of a man's encounter with a colossal whirlpool which *almost* eats him up but then spits him out, shows Bonaparte that 'the return to the womb has all the immensity of a vertiginous plunge into the ocean's yawning chasm'.[31] In 'The Masque of the Red Death', those noble-folk who have escaped the pestilence of the title which is ravaging the land lock themselves up with Prince Prospero in his castle, only to be 'found out' by the disease (in the form of a mysterious figure who enters and infects them). For Bonaparte,

> The appearance of *The Red Death* in Prince Prospero's palace, intended to represent the invading epidemic, is depicted by the entrance of a masked, blood-splattered, human form which strikingly and visually, characterizes the plague's symptoms.[32]

One problem which has been identified with this approach, as I indicated above, is the way in which this rigorous adherence to the model of the dream-work presents the text as a tissue of condensed and displaced images of real life, with that biographical life standing finally as the truth of the text. Another, as we shall see later in this chapter, is the way in which Freud's account of these primary processes then becomes the principle analytic tool, which is strictly applied to the text-as-dream. Nevertheless, Bonaparte's wonderfully suggestive encounter with Poe does something else too, which has more in common with subsequent work than has been generally noticed. It is Bonaparte's very involvement with the text, her meticulously close reading and the unconscious fantasy which she lends to it, which constitutes her interest for Elizabeth Wright. The Poe–Bonaparte contract thus becomes a protracted act of literary critical counter-transference, with critic-as-analyst infusing the interpretative situation with the colour of her own fantasy. For Wright, Bonaparte

> has shown [Poe] to be a master of ambivalence, and, above all, a provoker of fantasy, but she gives all the credit to his unconscious, whereas her unconscious too is actively involved. Though she believed herself to be the analyst of the text, she was as much an analysand as her patient.[33]

But I also now want to suggest a more intertextual way of situating Bonaparte's reading, since the intriguing collision of two texts which she directs need not necessarily be read as the mastery of one over another. Here Poe's Tales meet *The Interpretation of Dreams*, and though the latter is ostensibly the framing device and tool brought to bear upon the former in order to 'solve' it, Bonaparte is also making Freud's 'primary' text subject to the weird concerns of Poe. It remains the case that *The*

31 Ibid, p. 646.
32 Ibid, p. 646.
33 *Psychoanalytic Criticism*, p. 44.

Interpretation of Dreams is not overtly *about* literary texts, and the very fact that Bonaparte creates a site upon which its terms can be extended into the literary must change the way in which we read *The Interpretation of Dreams* itself. Once the channel is opened up, it cannot be made to flow in only one direction, from Freud *to* Poe. Through Bonaparte, Poe begins to show how the literary is already at stake in the most basic terms of psychoanalysis. Something of a cross-fertilization has begun to take place here – what Shoshana Felman is to call the 'interimplication' of each text in the terms of the other is already at work, even in this most 'applied' of readings.

II
Repeated Poe: Lacan and 'The Purloined Letter'

> [T]ruth manifests itself in the letter rather than the spirit, that is, in the way things are actually said rather than in the intended meaning. Literary critics learn how to read the letter of the text, how to interpret the style, the form, rather than just reading for content, for ideas. The psychoanalyst learns to listen not so much to her patient's main point as to odd marginal moments, slips of the tongue, unintended disclosures. Freud formulated this psychoanalytic method, but Lacan has generalized it into a way of receiving all discourse, not just the analysand's.[34]

If Marie Bonaparte takes on a range of Poe stories to produce a whole picture of a divided Poe, more recent poststructuralist psychoanalytic and deconstructive criticism has returned repeatedly to one short tale. Like Freud's 'Dora' case, the debate around 'The Purloined Letter' has been so prolific as to necessitate a substantial volume collecting the range of essays upon the subject.[35] Whilst 'The Purloined Letter' debate is only one element in a compound of psychoanalytic responses to Poe, its importance has extended beyond the specificity of Poe criticism. Unlike the gothic-macabre tales on which Bonaparte and others focus for their main psychoanalytic interest, 'The Purloined Letter' is strictly speaking a detective story, centring on Poe's 'serial detective' C. Auguste Dupin,

34 Jane Gallop, *Reading Lacan*, (Ithaca, NY, and London: Cornell University Press, 1985), p. 22.

35 See *The Purloined Poe*, ed. Muller and Richardson. This contains Lacan's 'Seminar on "The Purloined Letter"', Derrida's 'The Purveyor of Truth', and a selection from Marie Bonaparte, as well as the essays by Barbara Johnson and Shoshana Felman discussed below. The volume of papers collected concerning Freud's 'Dora' case has a similar purpose and organization, although most of the essays included are recent feminist responses, whereas *The Purloined Poe* evidences a debate which has been going on for much of this century.

who also appears in 'The Murders in the Rue Morgue' and 'The Mystery of Marie Rôget'. Poe has been identified as a crucial figure in the genesis of the detective story as well as 'originator' of both horror writing and science fiction, narrative forms which all in different ways focus equally on questions of meaning and processes of interpretation. And just as Poe is writing psychoanalytic detective stories, the analyst is also increasingly read *by* psychoanalysis as a detective. Lacanian critic Slavoj Žižek traces the convergence of analysis and detective-work:

> The "wolf-man," Freud's most famous patient, reports in his memoirs that Freud was a regular and careful reader of the Sherlock Holmes stories, not for distraction but precisely on account of the parallel between the respective procedures of the detective and the analyst . . . to dot the i's. It should be remembered that Lacan's *Écrits* begins with a detailed analysis of E. A. Poe's "Purloined Letter," one of the "archetypes" of the detective story, in which Lacan's accent is on the parallel between the subjective position of Auguste Dupin – Poe's amateur detective – and that of the analyst.[36]

It is, then, not only 'The Purloined Letter' which occupies us, but the thick complex of readings which have formed around it. One question we immediately face, raised by Barbara Johnson in her brilliant intervention in the debate ('The Frame of Reference: Poe, Lacan, Derrida'[37]), is that of paraphrase, of how critically to address not just the story itself (or any of the narratives on which I focus in this volume) but the accounts of readers of the story, through paraphrase, quotation, summary. A question of repetition is at stake here, which asks what happens when we repeat, as well as *how* we repeat what we repeat in analysis and writing. That repetition is a central psychoanalytic concern will become clearer as this study proceeds – repetition is not only a behavioural phenomenon (in the patient) but also a key narrative issue (the telling and retelling of stories in literary as well as psychoanalysis). If we are to follow through the analogy of text/reader as patient/analyst, we must question what is included and excluded from every critical repetition of the original 'story': how do texts tell only as much as they want to, only to be read for what lies between the lines, the implicit significance which exceeds or even contradicts the bare narrative, and in what ways do critical readers do the same? With the case of Poe, the debates seem to circulate around a series of interconnected repetitions, as critics keep returning to the repetitions of the story itself, and replay these moves in theoretical analysis. This also is the moment at which the interpenetration of literature and

36 Slavoj Žižek, 'The Detective and the Analyst', *Literature and Psychology*, 36/4 (1990), p. 29

37 'The Frame of Reference: Poe, Lacan, Derrida', in *The Critical Difference: Essays in the Contemporary Rhetoric of Reading* (Baltimore, MD: Johns Hopkins University Press, 1980), pp. 110–46. This essay is also collected in *The Purloined Poe.*

psychoanalysis begins to be more overtly debated, and the way in which repetitions in the story and the repetitions of critical analysis connect forms one focus for this. Johnson consequently asks the question of what it is to take and retell the story, to appropriate critically its bare bones, to remember and repeat it in a series of narrative returns. What is psychoanalytically at stake in critical repetition?

This question is significant here, since I too am faced with the problem of how to render the story in brief in order that I might then clearly demonstrate the ways in which it has been used. Readers embarking on analysis usually start with a scant plot synopsis, which acts as a bedrock of narrative co-ordinates upon which an analysis is then mapped. Johnson's concern is with what is at stake in this procedure – how critical appropriation of the elements of a story serve a psychoanalytic purpose. After offering her own selective paraphrase, Johnson then goes on to quote at length Lacan's own paraphrase of the plot, an initial moment of repetition in a series of concentric ripples with the 'original' story ostensibly at the heart of the pattern. (One of Johnson's concerns, as we shall see, is how this serves Lacan's double project of 'putting into' the story something which isn't strictly there, and leaving out of it something which might be.)

Yet I must nevertheless offer some context. The story concerns the quest to retrieve a letter received *by* and then stolen *from* the Queen, the potentially compromising contents of which are veiled. The King enters her boudoir; in order to conceal the letter from him, the Queen places it on a table to make it appear innocent. Minister D— enters and, perceiving the letter's importance, substitutes another for it, purloining the original and attaining a political power over the Queen by virtue of his possession, a power based in 'the robber's knowledge of the loser's knowledge of the robber'. This is the story's 'primal scene'. The Queen must retrieve the letter in order to be released from this compromising position, and so the Prefect of Police is called in. In secret, his officers microscopically scour the Minister's apartment, and when they fail to locate the letter, the detective Dupin is engaged. Reading the method of appropriation as a key to the method of concealment, Dupin surmises that the best way of hiding the object is to place it on show: 'Do you not see [the Prefect] has taken it for granted that *all* men proceed to conceal a letter . . . in some out-of-the-way hole or corner. . . . such *recherchés* nooks for concealment are adapted only for ordinary occasions.'[38] Dupin looks elsewhere – not to a hiding-place, but to the open room, and to what lies right in front of him. He visits the Minister, and spies the letter hanging visibly from a rack on the mantlepiece. Repeating the Minister's initial act of appropriation, Dupin substitutes another in its place, and is able to return the original to the Queen – a scene which repeats the first 'primal'

38 Edgar Allan Poe, 'The Purloined Letter', in *Selected Writings*, p. 341.

scene but with a key difference, for second time around a solution is reached.

Jacques Lacan's 'Seminar on *The Purloined Letter*' formed part of a series he presented in 1955 on Freud's *Beyond the Pleasure Principle*, entitled 'The Ego in the Theory of Freud and in the Technique of Psychoanalysis' (it was reworked to take the form now known as the opening essay of the 1966 *Écrits*). *Beyond the Pleasure Principle*, published just after the First World War, is best known as Freud's primary articulation of the theory of the death drive – a complex of ideas attempting to explain certain puzzling psychic phenomena which didn't 'fit into' Freud's earlier accounts of desire and mental regulation. Sadism and masochism, the pleasure of sexual extinction in 'the little death' of orgasm, the dynamics and patterns of a psychic trajectory which culminates in death as a goal, the repeated pleasures of fantasies and games focusing on loss (like the child's *fort-da* game discussed in *Beyond the Pleasure Principle*) – these phenomena could only be explained with reference to a new model of psychic economy and drives, since they did not accord with an earlier sense that sexual desire was the founding life instinct at the root of all psychic functioning, and that psychic life was dominated by the search for pleasure and the avoidance of unpleasure. Freud was, then, looking for a theory which went beyond the explanation of the pleasure/unpleasure principle as the governing economy of psychic life (hence, *Beyond the Pleasure Principle*). The theory of the death drive focused on the desire for pain as 'beyond' the desire for pleasure in several forms, and other essays, in particular 'The Economic Problem of Masochism' (1924), looked specifically at its various sexual manifestations. At the heart of masochistic practices lay both the destructive, suffering principle of the death drive and its other manifestation in psychoanalysis, the 'Nirvana principle', or the experience of death as a 'letting go', a moment of nullity or blissful absence. In this context, Freud evokes death as the 'aim' of life, in that it is the most exquisite moment of release, where energy and tension are discharged to allow the subject (itself extinguished at this moment) to dissolve into a zero-point of excitation, a moment of Nirvana, nullity, submission of the ego to the experience of its own loss.

But why pain or submission should be pleasurable for the masochist is only one question to which the death drive could potentially provide an answer. Freud was also faced with the phenomenon known as 'repetition compulsion'; and in its role as a response to *Beyond the Pleasure Principle*, Lacan's essay on 'The Purloined Letter' forms an extended meditation upon 'repetition compulsion'. For Freud, pain, and the possibility of pleasure in pain, was crucial to repetition compulsion. One of the factors which led him to the formulation of the death drive was the phenomenon whereby those who had experienced extreme trauma (for instance, soldiers in the First World War) continued to relive the trauma, with no apparent resolution, in fantasy, long after the original moment had passed.[39] Lacan notes that the germ of this idea is present in Freud's work right from the 1895 essay 'Project for a Scientific Psychology', in the idea that the unconscious (a concept Freud was then developing with the term *psi*) is 'caught up in the effort to find an irretrievable lost object'.[40] This is discussed by Muller and Richardson in *The Purloined Poe*: 'This movement takes the form not of a reminiscence of that object but of some kind of repetition (unconscious, to be sure) of the losing of it.'[41]

For Lacan, the question of repetition is focused more on loss than on pain (which is Freud's emphasis), and 'The Purloined Letter', with its repeated losses of a circulating object (first by the Queen, and then by the Minister), becomes in Lacan's reading a crucial literary articulation of this 'compulsion'. At the heart of Poe's 'original' is a repetition: it is a narrative within a narrative, as the story is itself told (repeated) by the 'I'-narrator who hears and repeats the tale of the filched letter, which is retrieved through the same pattern of the original filching. The original repetition is then critically repeated *ad infinitum* (my earlier narrative paraphrase, and this analysis of the story's analysts, is another example). So at the heart of the story itself lies both an absence and a pattern of repetition, which is then mimicked in subsequent readings which try to find the text out. In the words of the collection cited earlier, Poe is himself already 'the Purloined Poe', culturally retrieved through the same strategies suggested by his own text (with my – and other – paraphrases standing in for his 'original').

The story's second scene – which repeats the story's 'primal scene' but with a difference – is also the key to how Lacan understands the text as allegory. This is a repetition with understanding: the narrative repeats as psychoanalysis does, reworking the knotty anxieties of psychic prehistory. Thus it is only with the second scene that the whole story emerges as an allegory of the psychoanalytic situation, with Dupont act-

39 In Chapter 1 I discussed Freud's discovery of the genesis of neurosis in fantasy, but now I am discussing a different category of neuroses, caused by *real* traumas in the real world. The forms of neurosis which I discussed earlier (with which Freud is mainly concerned) are generally known as 'psychoneuroses', and have their origin in the mental events of early life. He did, however, also present a category known as 'actual neuroses', which develop not from a past event but from a present trauma: Laplanche and Pontalis define actual neurosis as 'A type of neurosis which Freud distinguishes from psychoneurosis. . . . The origin of the actual neurosis is not to be found in infantile conflicts, but in the present. . . . The opposition between the actual neuroses and the psychoneuroses is essentially aetiological and pathogenic: the cause is definitely sexual in both these types of neurosis, but in the former case it must be sought in "a disorder of [the subject's] contemporary sexual life" and not in "important events in his past life"' (Laplanche and Pontalis, *The Language of Psycho-Analysis*, p. 10, quoting Freud's 'Heredity and the Aetiology of the Neuroses').

40 Quoted by Muller and Richardson in their 'Overview' of Lacan's essay, *The Purloined Poe*, p. 56.

41 Ibid.

ing as analyst in 'restoring' the Queen to herself by finding and returning the letter.[42] This is for Lacan an act akin to the analytic move which rids the patient of her symptom, having reworked the significance of earlier memory traces. But lest this sounds too much like a crude decoding of actors in the plot into actors in the analytic situation, two points of difference (which I shall examine later in this chapter) should be stressed, both of which are underlined by Shoshana Felman in her reading of Lacan's essay. First, the way in which Lacan handles Poe's narrative repetition-with-difference is important to our understanding of identity in his work. '[W]hereas Marie Bonaparte analyzes repetition as the insistence of identity,' writes Felman again,

> for Lacan, any possible insight into the reality of the unconscious is contingent upon a perception of repetition, not as a confirmation of identity, but as the insistence of the indelibility of a difference.[43]

Secondly, if Dupont is in effect the Queen's 'analyst' (or the 'analyst' of her dilemma), as with any other act of analysis for Lacan, this is not necessarily because of any *knowledge* he has (of the letter, or the actors in the primal scene). Dupin-as-analyst doesn't 'cure' the situation because he has a certain knowledge, but because of the position he occupies in the chain of repetition. This is Felman:

> By virtue of his occupying the third position – that is, the *locus* of the unconscious of the subject as a place of substitution of letter for letter (of signifier for signifier) – the analyst, through transference, allows at once for a repetition of the trauma, and for a symbolic substitution, and thus effects the drama's denouement.[44]

It is not, then what the analyst *knows* which facilitates the analysis, but his position as activator of a repetition of the original problem – his occupancy of a position which facilitates transference. Marie Bonaparte's technique was the opposite of this, based on 'cracking the code' of authorial desire. She also urges that analysis should not be the work of the author, who must ideally remain ignorant of (and somehow separate from) the processes and images buried in his text for fear of censoring their 'guilty, libidinal parts':

42 Of this, Shoshana Felman writes: 'In what sense, then, does the second scene in Poe's tale, while repeating the first scene, nonetheless differ from it? In the sense, precisely, that the second scene, through the repetition, allows for an understanding, for an analysis of the first. This analysis through repetition is to become, in Lacan's ingenious reading, no less than an *allegory of psychoanalysis*. The intervention of Dupin, who restores the letter to the queen, is thus compared, in Lacan's interpretation, to the intervention of the analyst, who rids the patient of the symptom' ('On Reading Poetry,' p.147; her italics).

43 Ibid., p. 148.

44 Ibid., p. 147; her italics.

> Of all the devices employed by the dreamwork, that of the *displacement of psychic intensities* . . . is the most freely used in the elaboration of works of art, doubtless because such displacement is generally dictated by the moral censor, which is more active in our waking thoughts than in sleep. The conceiving and writing of literary works are conscious activities, and the less the author guesses of the hidden themes in his works, the likelier are they to be truly creative.[45]

Analytic mastery, which keeps the writer from direct access to parts of himself and ensures that access takes place via the analyst, is thus necessitated by the analysand's (here, the author's) presumed desire to self-censor, were he to 'know' the truth of himself. Lacan reads the situation differently, however, and not through the direct application of masterful 'codes'. A suspicion of mastery and knowledge is, then, crucial to how we understand Lacan, and how Lacan suggests ways for the understanding of text – even, or perhaps especially, *difficult* texts. Dupin then approaches the problem of how to find the letter as Lacanian analysis addresses a hard analytic situation, and as the Lacanian *reader* should approach an obscure or 'ungiving' text – not through prior knowledge and the desire to master, conquer, *solve* the mystery through power, but by entering into the mystery's own codes and strategies, opening it up *from within* using the tools *it* provides.

III
The Circulating Letter

I have said that it is the loss within the story, and the circulation of the lost object as its loss is replayed, which is fundamental to Lacan's reading of Poe through *Beyond the Pleasure Principle* – movement and circulation (rather than content and truth) are then key terms. Lacan's project, is generally speaking, to rework Freudian phases of human development from infancy onwards in the light of the structuralist linguistic developments of Ferdinand de Saussure and Roman Jakobson, emphasizing the interconnection of the subject's entry into language and the development of gender identification.[46] Lacanian psychoanalysis is consequently sometimes known as 'structural psychoanalysis', and its critical focus looks not to authorial symptoms breaking through on the page (*Sons and Lovers*

45 *Poe*, p. 645.

46 'Malcom Bowie tells us: "Lacan reads Freud. This is the simplest and most important thing about him" (1979, 116). But Lacan reads Freud (in German) in a manner shaped by the structural linguistics of Saussure and Jakobson, who afford a framework whereby Lacan can assert that "the unconscious is structured in the most radical way like a language"' (Muller and Richardson, Preface to *The Purloined Poe*, pp. x–xi).

as the expression of D. H. Lawrence's neurosis writ large) or the behaviour of characters (Paul Morel as Oedipal sufferer), but to the circulation of signs in the text.

Lacan's importance for literary criticism is, then, not simply that he offers a superannuated theory of gender and subjectivity – he brings a theory of language to (his reading of) key Freudian concepts of sexuality and psyche. Reworking the basic Freudian models of Oedipal development and castration anxiety, he maps out differences within the routes through which girls and boys enter language (the Symbolic order). It is this different linguistic relationship which underpins the child's uneasy entry into sexual difference at the same time. The child moves into the Symbolic Order as from an essentially narcissistic space (the Imaginary), characterized by the inability to see the world as Other. In the Imaginary, into which the child enters through the mirror stage, everything is an image of the child's own self. Sameness rather than difference means that here the child's experience is characterized by the illusion of unity – its own, and its own with the world. In moving forward into the Symbolic Order, the child has to confront Otherness in a number of forms (lack, castration, desire), as he recognizes his difference from a world (and a linguistic order) which he has not created, and which imposes on him the laws of the social order – he is, then, *subject to* a system he cannot control, and so his developing subjectivity comes into being through an experience of power (of the Other in its various forms) and *lack* (of control, of his own narcissistic self-unity which the social order does not recognize). Lacan writes in his famous essay on 'The Mirror Stage' (which I will look at again presently) that the end of this moment ushers the child through into the socially-governed system of the Symbolic:

> This moment in which the mirror-stage comes to an end inaugurates, by the identification with the *imago* of the counterpart and the drama of primordial jealousy . . ., the dialectic that will henceforth link the *I* to socially elaborated situations.[47]

It is here (when the 'I' is linked to 'socially elaborated situations') that desire (which is predicated on lack) becomes possible, and also here that entry into language begins uneasily to take place, at the same time as an equally difficult move towards gender. This, then, is the beginning of subjectivity, since it is only through language as a system of difference that subjective identity can be stated, thought, articulated – it is also the moment at which the unconscious is born, since here repression takes place. Sexual difference and language are thus both effects of entry into the Symbolic Order.

47 'The Mirror Stage as Formative of the Function of the I as Revealed in Psychoanalytic Experience', in *Écrits: A Selection*, p. 5; his italics.

This is a move which takes place as the child also learns of its separation from the mother, and begins to develop a relationship with the adult cultural order. For the boy, this is more straightforward. In the classic Freudian trajectory, after breaking with pre-Oedipal unity with the mother (which is followed by the Lacanian Imaginary, in which separation from external objects, such as the mother's body, is not recognized by the infant, which figures the mother as entirely a reflection of the child's narcissism), the boy-child moves through Oedipal attachment to the mother towards a recognition of the threat of castration represented by the father. He recognizes not only that the mother is not his to desire and have, but that she already 'belongs' to the father, and moreover that the father wields the power (Corman's *Pit and the Pendulum* is, as we have seen, a twisted image of this). In conjunction with this, the boy's visual image of the girl as 'castrated' (her different genitals) reinforces his fear of the father's power: if the girl has been damaged for her taboo desire, the boy could be too. Thus he abandons his Oedipal attachment, aligns himself with the father and henceforth 'loves' the mother only through acceptable substitute objects. Both this identification with the cultural 'Law of the Father' and the child's ability to substitute one thing for another thing (a symbol of mother – a substitute – for the mother one is not allowed to actually 'have', which one has lost), facilitates entry into a linguistic order within which words are also only substitutes, insignificant in themselves, gaining meaning only in their relationship with other words as part of a whole signifying system. For Lacan, full entry into language depends on apprehension of loss (of the mother).[48] As Ellie Ragland-Sullivan puts it,

> [F]or symbols or things to become language – to be verbally represented – they must first disappear as object or image. . . . Symbols only enter the field of language, making the *infans* (without speech) a speaking creature, when loss is added, when a symbol – such as a breast or a bottle – disappears as a fullness, when an infant loses the primary object, the mother as a presence. . . . Reading Freud on Identification Lacan was the first to realize that no seemingly coherent identification with language and culture could occur without a turning away from imaginary objects of plenitude (usually the mother) in the name of a Symbolic differential.[49]

48 As Jane Gallop puts it, this means that we are all castrated: 'Lacan's theory of sexual identification is precisely a theory of inadequacy, a theory of castration . . . But castration for Lacan is not only sexual; more important, it is also linguistic: we are inevitably bereft of any masterful understanding of language, and can only signify ourselves in a symbolic system that we do not command, that, rather, commands us. For women, Lacan's message that everyone, regardless of his or her organs, is "castrated," represents not a loss but a gain' (*Reading Lacan*, p. 20).

49 'The Symbolic', in *Feminism and Psychoanalysis: A Critical Dictionary*, ed. Elizabeth Wright (Oxford: Blackwell, 1992), p. 421.

Entry into language is differently negotiated by girls and boys, however. Girls do not view the mother, castration, or the threat of the father in the same way, in Freud's final model (from the early 1930s essays on femininity onwards), or in Lacan's work. They do not feel the loss of the mother's body as boys do, and so the work of symbolic substitution (and consequently their entry into the Symbolic) does not, for Lacan, come as easily. As Ragland-Sullivan again puts it, 'Language itself serves as the signified that tells the particular story of the knotting (or not) of the three orders (Imaginary, Symbolic, Real) in an individual's life in terms of acquisition of gender as an identity';[50] but it is a story which is told differently for girls and boys.

How, then, is this manifested in Lacan's reading of Poe? First, in his emphasis on the function of the letter itself. I have said that at the core of the story is not just a loss (of the letter) but an absence (of its content), as the details of the letter – what it actually *says* – are evaded through paraphrase. This is characteristically Poe – even his gothic tales tend to circulate around an absent centre or object, something either never revealed or revealed as a cipher. The revellers in 'The Masque of the Red Death' seize the mysterious figure who has penetrated their siege (which Bonaparte reads as an example of the dream-work's 'consideration of representability'), but he disappears as fast as they grasp him (he is, of course, the figure of their own deaths). After so much prevarication and ambiguity, Poe teeters on the brink of offering up the sign's *content*, the meaning or identity of the apparition, but when he gives us anything it is an absence:

> Then, summoning up the wild courage of despair, a throng of the revellers at once threw themselves into the black apartment, and, seizing the mummer, whose tall figure stood erect and motionless within the shadow of the ebony clock, gasped in unutterable horror at finding the grave cerements and corpse-like mask which they handled with so violent a rudeness, untenanted by any tangible form.[51]

Similarly, 'A Descent into the Maelström' offers an experience of pure, abyssal chaos, as the narrator is sucked into an enormous whirlpool which is nothing more than a *process*, the function of pure motion, as water is rendered a reflecting, rapid funnel into nothing.[52] Even 'Ligeia' is more concerned with the transformation of one woman into another,

50 Ibid.

51 *Selected Writings*, p. 259.

52 'The boat appeared to be hanging, as if by magic, midway down, upon the interior surface of a funnel vast in circumference, prodigious in depth, and whose perfectly smooth sides might have been mistaken for ebony, but for the bewildering rapidity with which they spun around, and for the gleaming and ghastly radiance they shot forth, as the rays of the full moon . . . streamed in a flood of golden glory along the black walls, and far away down into the inmost recesses of the abyss' (Poe, 'A Descent into the Maelström', in *Selected Writings*, p. 238.)

changing forwards and back, than with the fixed identities of either Ligeia or Rowena.

It is this emphasis on process rather than on revelation or substance which is Lacan's focus here – like the figure of the Red Death, Lacan's signs are also 'untenanted'. We hear nothing of the purloined letter's content: it is passed around, at each turn demonstrating something of the relationship between people, and in this sense it acts as the means by which relationships are established. Glossing Lacan, Muller and Richardson put it this way:

> As the letter passes from the Queen to the Minister to Dupin to the Prefect back to the Queen, the content remains irrelevant, and the shifting parameters of power for the subjects concerned derive from the different places where the letter is diverted along this "symbolic circuit" . . . [T]he "place" of the signifier is determined by the symbolic system within which it is constantly displaced.[53]

For Lacan, this makes the letter itself a 'pure signifier' because it has been completely estranged from its signified: what the letter *means* is entirely irrelevant to the way in which it is passed around in the story. The story plays out a pattern,

> in which the subjects relay each other in their displacement during the intersubjective repetition.
>
> We shall see that their displacement is determined by the place which a pure signifier – the purloined letter – comes to occupy in their trio.[54]

The act of repetition is then divorced from the meaning or content of what is repeated. For Lacan, the letter functions in the story not because of what it *says* and *is* in terms of its content (which we never know), but because of its *position* between people, its *role* in producing certain interpersonal effects and subjective responses: the Queen's embarassment and fear, the Minister's power in its possession, Dupin's act of retrieval understood as a moment of revenge carried out through the very same strategy which was used to take it in the first place – even the King's ignorance of the letter's very existence. All of these moves focus on the letter, a letter which forces response without ever revealing itself. Each character finds him- or herself already in a relationship of action and reaction with a sign which has no essential presence – it is never 'filled out':

> [T]he letter was able to produce its effects *within* the story: on the actors in the tale, including the narrator, as well as *outside* the story: on us, the readers, and

53 'Lacan's Seminar on "The Purloined Letter": Overview', in *The Purloined Poe*, pp. 57–8.
54 'Seminar on "The Purloined Letter" ', in *The Purloined Poe*, p. 32.

> also on its author, without anyone's ever bothering to worry about what it *meant*.[55]

It is because we never know its content (or its origin[56]) that its function, as it circulates between people, can be revealed all the more starkly. Each subject in the story is changed by the effects of the letter-as-signifier, an instance in miniature of how the subject does not create and control, but is constituted by and of the Symbolic Order.

The subject is, then, constructed through a crucial developmental relationship to the sign as an image of lack. For Lacan, the combination of absence and movement which characterize the letter make it the exemplary signifier: it is both there and not there – *not* there in the sense that it lacks an identifiably meaningful content. It is there but it is not detected, it is there but its content is not, it is there but we know nothing of it. It behaves in the story as signs do in psychic life, since its 'heart' is always absent: 'the signifier is a unit in its very uniqueness,' writes Lacan, 'being by nature symbol only of an absence':[57]

> [W]e cannot say of the purloined letter that, like other objects, it must be *or* not be in a particular place but that unlike them it will be *and* not be where it is, wherever it goes.[58]

This is also, then, a story centrally concerned with things out of place but never wholly lost, with the significantly misplaced but apparently insignificant detail. In this sense it could also serve as a statement of the Freudian unconscious in which nothing is lost (to which I will return in Chapter 4, in a discussion of the 'Wolf Man'). In his 'Note Upon the "Mystic Writing-Pad"' (1925 [1924]) Freud discusses a curious apparatus (the device of the title, a writing-tablet which 'provide[s] an ever-ready receptive surface [for writing] and permanent traces of the notes that have been made upon it [and erased]'[59]). In this little invention Freud finds a model for the layering of conscious perception and unconscious memory, in which lies the trace of all experience but also its apparent erasure by the existence of current (conscious) preoccupations. 'I do not think it is too far fetched', he writes,

> to compare the celluloid and waxed paper cover with the system Pcpt.-Cs. and its protective shield, the wax slab and the unconscious behind them, and the appearance and disappearance of the writing with the flickering-up and passing-away of consciousness in the process of perception.[60]

55 Quoted and translated by Johnson in *The Critical Difference*, p. 115.
56 'The tale leaves us in virtually total ignorance of the sender, no less than of the contents, of the letter' (Lacan; 'Seminar on "The Purloined Letter"', p. 41).
57 Ibid., p. 39.
58 Ibid.
59 Freud, 'A Note Upon the "Mystic Writing-Pad"' (1925), PF vol. xi, p. 431.
60 Ibid., p. 433.

Earlier in the essay, Freud had discussed human memory as having 'an unlimited receptive capacity for new perceptions and nevertheless lay[ing] down permanent – even though not unalterable – memory-traces of them'.[61] Nothing, then, is lost, although it may not all be immediately present. But these buried traces do not 'flicker up' on demand, and neither could you discern them by looking for them with any straightforward gaze. Like the Mystic Writing-Pad, the fabric of Poe's story conceals, as if to lose forever, the object, which in fact remains only buried.

In this sense, too, the letter offers a focus for an allegory of gazes – a story concerning the 'right way of seeing', for only those who know how to look for it will find it, only a specific interpretative strategy will reveal the object (even though its content remains veiled). The police, failing to find the letter, are 'realists', operating on reality with a common-sense stupidity which can only see selectively. Theirs

> is the realist's imbecility, which does not pause to observe that nothing, however deep in the bowels of the earth a hand may seek to ensconce it, will ever be hidden there, since another hand can always retrieve it, and that what is hidden is never but what is missing from its place, as the call slip puts it when speaking of a volume lost in a library. And even if the book may be on an adjacent shelf or in the next slot, it would be hidden there, however visibly it may appear.[62]

To Dupin, as to the psychoanalyst, nothing in the unconscious (or the library, or the text) is ever lost or finally hidden. Critical and psychoanalysis, like detective work, thus becomes a question of developing the right eye, of casting off the 'imbecility' of the realist.

But what is the relationship of this semiotic analysis with the questions of desire and subjectivity which have been crucial to the forms of psychoanalysis we have encountered so far? For Lacan desire itself is predicated on absence, and the development of subjectivity takes place in terms dictated by the sign. 'The Purloined Letter' thus demonstrates how crucially subjectivity and the Symbolic are linked for Lacan: '[I]t is the symbolic order which is constitutive for the subject – by demonstrating in the story the decisive orientation which the subject receives from the itinerary of the signifier.'[63] And it is not only the subject which is thus orientated; it is this skewing which, for Lacan, 'makes the very existence of fiction possible'.[64]

This pattern of repetition which takes place both inside and outside the

61 Ibid., p. 430.
62 Lacan in *The Purloined Poe*, p. 40.
63 Ibid., p. 29.
64 Ibid.

story is, then, like the processes of the psychoanalytic situation itself, with the text (or dream, or other textual unconscious manifestation) as focus, rather than the author or character's psychic state. One approaches texts, whether literary or analytic, by following 'the path of the signifier';[65] as Bice Benvenuto and Roger Kennedy put it,

> The Poe essay is a reasonably clear illustration of Lacan's notion of the Symbolic Order in that he uncovered what he considered to be similarities between the story and the psychoanalytic situation, for example concerning the kind of knowledge needed to discover the patient's truth. In analysis a 'letter' can be found, put aside, diverted or hidden by the patient. The basic analytic task, in Lacan's view, was to find this letter, or at least find out where it is going, and to do this entails an understanding of the Symbolic Order.[66]

The letter then acts as the key 'decentring' element in the text, in that it 'insists' on being 'heard' and affecting everyone in its proximity, even if they cannot decode its meaning into a consciously realized truth. In this sense it exemplifies what Lacan calls 'the insistence of the signifying chain', the chain of unconscious thoughts which Freud identified as the controlling psychological element.

IV
Lacan, the Gaze and Film Theory

Another issue which runs through Lacan's text is the question of the gaze, which is central to his theorization of the development of the subject and, in turn, crucial to the important work on subjectivity and the image which took its cue from Lacan, in feminism and film theory from the 1970s onwards. I have said that 'The Purloined Letter' for Lacan hinges on a dynamic of *mis*-seeing; it also offers an allegory of how the gaze operates in psychic life. Lacan's seminal essay of 1949, 'The Mirror Stage as Formative of the Function of the I', is often seen as the starting-point for this work, focusing upon the moment at which the infant (aged between six and eighteen months) first identifies with a whole 'gestalt' image of itself as separate from the world, a unity which is self-sufficient and contained. No longer, from this moment onwards, undifferentiated from the mother's body or the sensual world around him, the child recognizes itself through identification of and with a self-image – a reflection in the mirror, or through identification with another body like its

65 Bice Benvenuto and Roger Kennedy, *The Works of Jacques Lacan: An Introduction* (London: Free Association Books, 1986), p. 102.
66 Ibid., p. 101.

own. This 'startling spectacle of the infant in front of the mirror' is described by Lacan thus:

> Unable as yet to walk, or even to stand up, and held tightly as he is by some support, human or artificial . . ., he nevertheless overcomes, in a flutter of jubilant activity, the obstructions of his support and, fixing his attitude in a slightly leaning-forward position, in order to hold it in his gaze, brings back an instantaneous aspect of the image.[67]

This is a transformative experience, signalling for the child entry into the Imaginary: the child imagines itself as a whole through identification with the image it sees. 'We have only to understand the mirror stage *as an identification*,' Lacan continues, 'namely, the transformation that takes place in the subject when he assumes an image.'[68] The Imaginary is then opened up not only by the narcissism of the child at this moment before 'lack' intrudes, but by an essentially visual experience (whether the child goes through this by actually looking in a mirror or not, still the moment is psychologically a visualization of the self as whole). Although this experience does not strictly herald the birth of the subject (which for Lacan comes with entry into the Symbolic), it nevertheless sets up what Laplanche and Pontalis call 'the first roughcast of the ego'.[69] A brief literary diversion might clarify this.

In her short story 'Flesh and the Mirror', Angela Carter offers an intricate literary reflection on this 'mirror phase', an experience of the dawning of the 'I' written from a moment afterwards, once the I has passed through into a later – cultural – moment. Returning to a foreign country she is beginning to call home, the narrator offers an (actual) encounter with a mirror which consolidates a sense of self which is unclear when the story opens:

> The magic mirror presented me with a hitherto unconsidered notion of myself as I. Without any intention of mine, I had been defined by the action reflected in the mirror. I beset me. I was the subject written on the sentence of the mirror.[70]

'I beset me' is a statement the subject can only make after the 'I' has begun to be linguistically constructed. The passage is in the past tense,

67 'The Mirror Stage', pp. 1–2.
68 Ibid., p. 2; his italics.
69 Laplanche and Pontalis, *The Language of Psycho-Analysis*, p. 251. In his introduction to the *New Left Review* publication of 'The Mirror Stage', Jean Roussel writes that 'we should understand that the subject is only constituted for the first time in this primary identification, which should therefore be regarded not as an identification in the proper sense but as what first makes identification possible' ('Introduction to Jacques Lacan', *New Left Review*, 51 (Sept.–Oct. 1968), p. 68).
70 In *Fireworks* (1974; London: Virago, 1987), pp. 64–5.

allowing for the fact that only after she has been written 'on the sentence of the mirror' can she speak of the event at all. Here, then, the mirror stage is consolidated as a passage through into coherent subjectivity and cultural identity – for Lacan this will take place later, with the intrusion of lack into the subject's psychic landscape, and the real beginning of subjectivity proper. Carter's narrator's mirror-image is both the ostensibly integrated 'I' she is from now on, and the life-sentence imposed by an entry into language. Only once the image of her identity has begun to be constructed and performed can she look back and rewrite the history of 'before'.

But I have not begun at the beginning: the mirror is seen some way into the story. This would imply that, as the story opens, it must represent a pre-subject, a non-self of the infantile times before the history of the 'true' subject gets under way. So how does one write of a self before the self, expressing the 'body in bits and pieces' before language (Lacan's *corps morcelé*), representing that which is there before representation? Where has the story begun? There is, indeed, an 'I' at the start of Carter's fable, but it is one which slips from first to third person without a by-your-leave to the confused reader ('however hard I looked for the one I loved, she could not find him anywhere'[71]). Like Freud's 'His Majesty the Baby', she sees the world without edges as the narcissistic extension of her sovereign will ('And I moved through these expressionist perspectives in my black dress as though I was the creator of all and of myself . . . as though the world stretched out from my eye like spokes from a sensitized hub that galvanised all to life when I looked at it'[72]). There is an 'I' here, but it cannot truly differentiate itself through identification, through the assumption of a single, reflected image. The action of the mirror is a passage into another country: 'The bureaucracy of the mirror issues me with passport to the world.'[73]

Lacan's mirror phase is often understood as a transformative moment in the construction of the subject, at which it is constituted in terms of the image of itself which the infant sees before it in the mirror (in this sense, Lacan's *stade du miroir* is better translated as mirror *phase*, since it represents a turning-point in the development of the infant, rather than a distinct period[74]). Carter's is a fantastic adult fabrication – indeed, a fictive critical analysis – of the formative experience, woven to show the process through which a quite different subject (a woman in control of the gaze) might be also constructed through the looking glass, for the world on *this* side of the mirror is a world within which she finds herself visually dominant. The female narrator of 'Flesh and the Mirror' has a lover made in the image of her fantasy, a reversal of the drama in the later story, 'The

71 Ibid., p. 63.
72 Ibid., p. 62.
73 Ibid., p. 65.
74 Laplanche and Pontalis, *The Language of Psycho-Analysis*, p. 252.

Snow Child', in which a man conjures up a fantasy-girl purely from his own desire, who lives only as long as does his desire. Like much of Carter's work, 'Flesh and the Mirror' is a site upon which visual identification and the liberation of fantasy collide in the service of feminism. Carter is known (and is notorious[75]) for her attempt to turn the world of visual relations upside down, making men the objects and voracious women their sexual consumers (she continued to be interested in this right to the end of her writing, as we shall see in the next chapter). 'Flesh and the Mirror' is perhaps the starkest example of this: Of the lover's face she writes:

> I suppose I do not know how he really looked and, in fact, I suppose I shall never know, now, for he was plainly an object created in the mode of fantasy. His image was already present somewhere in my head and I was seeking to discover it in actuality . . . his self, and by his self, I mean the thing he was to himself, was quite unknown to me. I created him solely in relation to myself, like a work of romantic art, an object corresponding to the ghost inside me.[76]

As the pure reflection of her own narcissistic image, 'he' is not in any sense other or different – this is a fantasy still played out in the mirror, an experience which is both Imaginary and Symbolic, mediated by the 'I' and yet unmediated in its total identification which makes the 'he' only an image of the seeing self. This gazing woman works through a drama within which first the subject is constituted ('written on the sentence of the mirror') through a visual experience, but then this experience donates the power to infect and affect the external world, made in the image of (female) desire (a fantasy – of course!). Carter's story is also a good example of literature turning back to psychoanalysis, not for a set of applicable, 'given' theories through which it *can be read,* but as an alternative source of narratives through which it *will read itself.* As a knowing employer of diverse cultural registers, Carter refuses to be 'faithful' to only *one* (literary) disciplinary source. Thus both 'The Mirror Stage' and Carter's fictional re-reading of it offer dramas of the subject beginning to be constituted through a visual experience.

This extends through Carter's work in her repeated return to the images of classic Hollywood as source of visual and narrative motifs, as well as a space in which the sexual subject is made up and unmade (I shall return to this again towards the end of Chapter 3). Taking this influence in another direction, the mirror phase as a theory of visual identifi-

75 I am here thinking particularly of responses to the macho hero of *Heroes and Villains,* which is discussed in the interview Lorna Sage carried out in 1977. Here Carter says, 'One of the things I was doing then unconsciously and am now doing consciously is describing men as objects of desire. I think a lot of the ambivalence of response I get is because I do this' (Sage, 'A Savage Sideshow', *New Review,* 4/39–40 (July 1977), p. 55).
76 pp. 67–8.

cation, the primal scene as a formative visual fantasy-moment, as well as other visually focused moments in Freud's account of fetishism (which I shall also look at in Chapter 3), have had a massive impact on film studies since the 1970s. There *are* examples of applied Freudian film analysis like Bonaparte's reading of Poe (psychobiographical readings of the director-*auteur* carried out through the body of his work), but the most widespread influence of psychoanalytic theory in film analysis is articulated through Lacan and Freud's models of the role of the image in the construction of the subject.[77] Here, theorists often slide between the work of Freud and Lacan, assuming the cinematic use of terms such as the Symbolic and the Imaginary, but working explicitly with Freudian models (such as scopophilia – comprising the drives to look and be seen, or voyeurism and exhibitionism) to emphasize how the audience responds to and receives film, how it is 'sutured' into its narratives techniques, and how the film-goer becomes ideologically constructed as a particular subject by the cinematic apparatus. The influence of Louis Althusser's essay 'Ideology and Ideological State Apparatus (Notes Towards an Investigation)' and Roland Barthes's semiotic analysis of the Oedipal narrative structure of a literary text in *S/Z* were tremendously important for the development of film theory in the 1970s; in particular, Jean-Louis Baudry's seminal essay, 'Ideological Effects of the Basic Cinematographic Apparatus' (first published in English in 1975), developed a way of reading film which actively used Lacan and Althusser together. This mixture of Althusserian marxism, semiotics, and Lacanian psychoanalysis characterized much of the distinctive work carried out under the aegis of the important film journal *Screen* from the mid-1970s to the early 1980s.

As is the case with literary analysis, psychoanalysis offers film theorists ways of accounting for cinematic representations in terms of models of how the subject is formed sexually, linguistically and socially. Psychoanalysis has been so useful to cultural critics precisely because it allows analysis of identity and representation, of the subject and the culture into which he or she emerges, to take place *together*. Feminist film theory in particular has opened up discussion of cinema and has developed theories of spectatorship (the subject constructed through his or her act of looking – as a psychological, sexual and cultural act) which gained a broad (and even popular) currency in the 1980s. The notion of 'the male

[77] E. Ann Kaplan makes the point that although the most dominant approach of film theory (influenced and characterized by the work of the journal *Screen* from the mid-1970s to the mid-1980s) was known as 'Lacanian film theory', this is 'an ironic labeling in that many Freudian concepts were central and only limited aspects of Lacanian thought were involved (ie., the mirror phase, the distinction between the Imaginary and the Symbolic, the notion of the unconscious as "structured like a language," and the constitution of the subject as "split" at the moment of entry into language, which is also entry into lack/desire)' (Introduction to *Psychoanalysis and Cinema* (New York and London: Routledge, 1990), p. 9).

gaze' which became widely used primarily derived from Laura Mulvey's highly influential essay, 'Visual Pleasure and Narrative Cinema', first published in *Screen* in 1975, but extensively anthologized ever since. Other important work from the mid-1970s which made use of psychoanalytic models include Claire Johnston's work on alternative feminist film practice and Pam Cook's textual analyses of contradictory images of women, whilst the two worked together on a crucial essay, 'The Place of Woman in the Cinema of Raoul Walsh'. Published a year before Mulvey's piece, this laid out a Lacanian agenda as a way of bypassing the 'symptomatological' or symbolic Freudian readings at which I have already looked; in many ways, the essay single-handedly brought feminism, film and Lacanian psychoanalysis together to form a new agenda. Cook and Johnson state their aims boldy but simply: 'the authors are attempting to help lay the foundations of a feminist film criticism as well as producing an analysis of a number of films directed by Walsh'.[78] They go on to argue that the ostensibly threatening Jane Russell character in *The Revolt of Mamie Stover* actually functions in the narrative to signify not potency and plenitude but woman as *lack*. Certainly, Russell as Stover *looks back* at men (a point taken up by Linda Williams more recently[79]), from a position loaded with threat:

> Her "look" – repeated many times during the film, directed towards men, and explicitly described at one point as "come hither" – doubly marks her as signifier of threat. In the absence of the male, the female might "take his place": at the moment of Jane Russell's "look" at the camera, the spectator is directly confronted with the image of that threat.[80]

Yet more important is Russell/Mamie's role as a circulating object (rather like the purloined letter – an argument made here with reference to Lévi-Strauss's structural-anthropological work on woman as object of exchange). She is a protagonist who 'cannot write her own story'; as a woman who functions as lack,

> she is a signifier, an object of exchange in a play of desire between the absent subject and object of the discourse. She remains "spoken": she does not speak.[81]

Laura Mulvey's article, which situates the female screen body in terms of an objectifying (voyeuristic and fetishistic) male gaze as marked by

78 Collected in *Feminist Film Theory*, ed. Constance Penley (London and New York: Routledge and BFI, 1988), p. 26.
79 'When the Woman Looks', in *Film Theory and Criticism*, ed. Gerald Mast, Marshall Cohen, and Leo Braudy (Oxford: Oxford University Press, 1992), pp. 561–77.
80 Cook and Johnston, 'The Place of Woman in the Cinema of Raoul Walsh', p. 31.
81 Ibid., p. 33.

castration and lack, took the implications of Lacan's thought, mediated by French theorist Christian Metz, in another direction (in many ways Mulvey's work here is a feminized reading of Metz's crucial work on cinematic scopophilia). 'Visual Pleasure and Narrative Cinema' assumes an understanding of the development of the subject through the Imaginary into the Symbolic, but focuses on how theories of voyeurism and fetishism position the spectator (the audience, camera, and male protagonist on screen). In Mulvey's view, mainstream Hollywood cinema is formed in the image of male desire, and its glamorous female stars are constructed as objects satisfying or filling the gap opened up in the male subject by castration anxiety. Whilst this has resonances with Cook and Johnston's reading of Jane Russell, Mulvey develops this by detailing what forms of perverse pleasure, what types of male gaze, are at work in constructing the female image in terms of what Mulvey calls woman's 'to-be-looked-at-ness'.[82] The female spectator, in Mulvey's, early view, has no active point of entry into a form constructed around a scopophilic desire which sees women only as objects of lack, with 'Woman as Image, Man as Bearer of the Look'.

Part of the importance of this work has been in the psychoanalytically informed debate concerning women's position in the cinematic apparatus, and their pleasures as a discrete audience in cinema history, which Mulvey's essay engendered. This debate has encompassed important alternative ways of figuring the female spectator in terms of possibilities of masquerade or alternative visual pleasures and positions (here works by Mary Ann Doane, Judith Mayne and Jackie Stacey are important),[83] as well as Mulvey's own later intervention through a revision of her original position. Many other – not specifically feminist – approaches to films

82 This term has had considerable currency, but it is taken up again in the 1980s in debates around the male spectacle, starting with Stephen Neale's important *Screen* essay, 'Masculinity as Spectacle', and Richard Dyer's 'Don't Look Now: The Male Pin-Up' (both collected in *The Sexual Subject*, ed. *Screen* (London and New York: Routledge, 1992). See also Peter Middleton's *The Inward Gaze: Masculinity and Subjectivity in Modern Culture* (London and New York: Routledge, 1992) and Rowena Chapman and Jonathan Rutherford, eds., *Male Order: Unwrapping Masculinity* (London: Lawrence & Wishart, 1988). For a discussion of this debate, as well as an account of how it might be usefully transferred back into work on the literary text, see Linda Ruth Williams, *Sex in the Head: Visions of Femininity and Film in D. H. Lawrence* (Hemel Hempstead: Harvester-Wheatsheaf, 1993), ch. 3.

83 See Mary Ann Doane's work on woman's viewing position as masquerade (which takes its cue from Joan Riviere's 'Womanliness as Masquerade'), and collections on the woman's look including *The Female Gaze* anthology, ed. Lorraine Gamman and Margaret Marshment (London: Women's Press, 1988), the special edition of *Camera Obscura* on *The Spectatrix*, 20–1 (1989), ed. Janet Bergstrom and Mary Ann Doane, and E. Deidre Pribram, ed., *Female Spectators* (London and New York: Verso, 1988). See also Laura Mulvey's 'Afterthoughts on "Visual Pleasure and Narrative Cinema" inspired by *Duel in the Sun*', in *Visual and Other Pleasures* (Bloomington: Indiana University Press, 1989).

navigated through various aspects of psychoanalytic theory have been developed, but it is not possible to detail each of them here.[84] Examples might include Peter Baxter's reading of *The Blue Angel* as dream, worked through a discussion of the dream-work's process of secondary revision (but here read in quite a different way from Bonaparte's work on Poe and the dream-work), or Bruce Kawin's reading of horror films as nightmares, or Kaja Silverman's alternative reading of the male spectator position through theories of male masochism.[85]

To return again to the literary, how do these questions of looking and identity manifest themselves in 'The Seminar on "The Purloined Letter"', or in 'The Purloined Letter' itself? A strong dynamic of power exchanged between the seeing and the blind operates in Poe's story. Concealment of the lost, guilty object is ensured not by making the thing itself invisible (hiding the letter in a secret drawer, say, or under the floorboards) but by understanding that 'concealment' can be ensured on quite different conditions: the story is 'about an observer being observed without observing that she is being watched in turn'.[86] Both acts of purloining the letter rely upon an understanding that the subject is blind in certain situations, even though he or she can technically see – what Freud was to call, in *Studies on Hysteria*, 'that blindness of the seeing eye'.[87] What Lacan refers to as the story's 'primal scene' – its establishing moment – is like all primal scenes in classic psychoanalysis, predicated upon a dynamic of seeing and unseeing – another key visual experience in the formation of the subject, which finds its way into Lacan's theory of gender alongside the mirror phase. From this scene in the story, the letter passes on the first stage of its journey.[88] The Queen places on display, as if to conceal it, the guilty object which threatens the law, the position of the King. Whilst the three characters here – King, Queen and Minister – cannot be resolved or decoded as the father/mother/child of the Freudian tryptich, for Lacan the visual power and anxieties of the primal scene are nevertheless present.

84 A good general reader is Kaplan (ed.), *Psychoanalysis and Cinema.*

85 Peter Baxter, 'On the Naked Thighs of Miss Dietrich', in *Movies and Methods*, ii, ed. Bill Nichols (which also contains other essays using psychoanalysis in film criticism, with useful introductions by Nichols); Bruce Kawin, 'The Mummy's Pool', in *Film Theory and Criticism: Introductory Readings*, ed. Mast *et al.*; and Kaja Silverman, *Male Subjectivity at the Margins* (New York and London: Routledge, 1992).

86 Muller and Richardson, Preface to *The Purloined Poe*, p. vii.

87 p. 181 n.

88 'Lacan compared the first scene to a primal scene, a scene of sexual intercourse between the parents which the child observes, or his phantasy of what he observes. Freud emphasized how the primal scene is grasped and interpreted by the child later, when he can put it into words; in Lacanian terms, when he can link the imaginary experience into the Symbolic Order' (Bice Benvenuto and Roger Kennedy, *The Works of Jacques Lacan* (London: Free Association Books, 1986), p. 94).

But more than this, the story is a web of exchanged glances. People see each other's actions, and *see that they are being seen*, the primary element in Lacan's theory of the gaze, which is always rooted in what is at stake in being observed: 'we are beings who are looked at, in the spectacle of the world,' writes Lacan in the *Four Fundamental Concepts of Psycho-Analysis*,[89] where he also tells a little story of a visit to the seaside. His companion, one Petit-Jean, points out a tin can floating on the water:

> It glittered in the sun. And Petit-Jean said to me – You see that can? Do you see it? Well, it doesn't see you!

Lacan is disturbed by this (although Petit-Jean finds it amusing), and analyses why:

> if what Petit-Jean said to me, namely, that the can did not see me, had any meaning, it was because in a sense, it was looking at me, all the same. It was looking at me at the level of the point of light, the point at which everything that looks at me is situated.[90]

The 'relation of the subject with the domain of vision' is then to be a screen, receiving the Other's gaze but lacking its plenitude, not the reverse. For Lacan, then, the subject does not possess the gaze, but is primarily constituted by it. This supremely paranoid concept is an important element of the child's growing experience of the world as Other, in its entry into the Symbolic. Indeed, this is already a component of the child's experience of its *gestalt* image in the mirror phase, for although I said that this was primarily a narcissistic experience (in which the child saw an image of itself in the mirror), it is also the first moment at which the child relates to himself as if his image were that of another. This moment is the beginning of difference, which is to be fully confronted as the Other which the child confronts upon entering the Symbolic.

Thus in reading 'The Purloined Letter', Lacan identifies a dynamic of glances which are more important for the way they act out a drama of being seen than of seeing – the power one might have only has meaning in terms of *being seen to have* that power by the other's glances. I have said that the letter, with its absent-content, has no identity of its own save the role it plays in being passed around. Nevertheless, the letter lends its possessor a certain power, and this is its role as an object of desire as it passes between people: 'The ascendancy which the Minister derives from the situation is thus not a function of the letter, but, whether he knows it

89 The Four Fundamental Concepts of Psycho-Analysis (1973), trans. Alan Sheridan (New York and London: Norton, 1981), p. 75.
90 Ibid., both quotations p. 95.

or not, of the role it constitutes for him.'[91] There is blindness in this story, but there are also at least two forms of seeing. The letter has power because it is seen to be possessed by some characters (whose visual positions change). Lacan's emphasis on repetition thus takes a visual turn, as he focuses 'three moments, structuring three glances, borne by three subjects, incarnated each time by different characters':

> The first is a glance that sees nothing: the King and the police.
> The second, a glance which sees that the first sees nothing and deludes itself as to the secrecy of what it hides: the Queen, then the Minister.
> The third sees that the first two glances leave what should be hidden exposed to whoever would seize it: the Minster, and finally Dupin.[92]

Here, there is visual exchange and recognition at each twist of the tale. Along with the power which the letter gives to its possessor, the power of the glance is also passed around – from Queen to Minister, from Minister to Dupin. The gaze, and a subject position founded in it, is thus something liable to shift as the subject's relationship to the others in its network also changes, but it is always set up in terms of the difficult gaze of the other rather than the potent look of the self.

If psychoanalytic visual theory has found a very wide currency in the contemporary film theory here discussed, it has also found a very specific life in the analysis of the circulation of power in a literary text. Here, then, Lacan's own theories of the gaze negotiate the visual dynamic represented in 'The Purloined Letter' itself, a move closer to applied psychoanalysis than we might initially have thought Lacan would be engaging in (although he does, of course, resist a specific analysis of author-through-text: 'where the Freudian method is ultimately biographical' writes E. Ann Kaplan, 'Lacan's is textual'). In the final section of this chapter I shall turn to this question of the priorities of literature in relation to psychoanalysis as a 'master discourse', of its strategies for resisting this role, and of the implications of 'applied psychoanalysis' itself, again with reference to 'The Purloined Letter'.

91 Lacan, 'Seminar on "The Purloined Letter"', p. 46.
92 Ibid., p. 32.

V
'The Unconscious of Psychoanalysis': Felman, Johnson and Applied Psychoanalysis

[I]n a reversal of what is "usually felt," we might rather consider psychoanalysis one application of literary studies.[93]

[I]n the same way that psychoanalysis points to the unconscious of literature, literature, in its turn, is the unconscious of psychoanalysis; that the unthought-out shadow in psychoanalytical theory is precisely its own involvement with literature . . .[94]

Like a number of American critics known internationally for their affinity with work carried out in the journal *Yale French Studies*, Barbara Johnson and Shoshana Felman are both engaged in a process which rethinks the relationship between psychoanalysis and literature, and strongly interrogates 'applied psychoanalysis' – a term often used to characterize the bald application of 'vulgar Freudianism' to the text, a text which is then in turn read by the vulgar Freudian as the transparent rendering of the author's unconscious drives and desires. Freud, indeed, has also been charged with a form of 'vulgar Freudianism' in his own essays on art and literature.[95] Felman's work in particular has problematized the classic power relationship which establishes literature as passive matter 'available' to demonstrate or fulfil the primary theoretical assertions of psychoanalysis. Thinking through what is at stake in the rich cross-fertilization of literature and psychoanalysis, Felman undoes this trajectory of application which posits 'theory' as the primary partner, practising its models on the literary text, which then becomes the secondary 'proof' of that theory's models.

But in deconstructing the operation of one body of thought upon another textual body, Felman is not concerned with reversing the balance of power, so that literature subjects psychoanalysis to its will instead of the other way around. Rather, Felman wants to make the whole way in which we understand how one body of thought 'reads' another open to question, and she does this through a series of startling close discussions

93 Gallop, *Reading Lacan*, p.25.

94 Shoshana Felman, 'To Open the Question', in *Literature and Psychoanalysis: The Question of Reading: Otherwise*, ed. Felman (Baltimore, MD, and London: Johns Hopkins University Press, 1980) p. 10.

95 Of these the general opinion has been, in Peter Brooks's words, that 'Freud speaks most pertinently to literary critics when he is not explicitly addressing art' ('The Idea of a Psychoanalytic Criticism', in Shlomith Rimmon-Kenan (ed.), *Discourse in Psychoanalysis and Literature* (London and New York: Methuen, 1987), p. 5.

which marry a sense of historical debate (the development of critical readings, say, of Poe or James, or the moves of psychoanalytic history) with an awareness of how texts operate on their readers. Felman is concerned to prioritize the literary in its widest as well as its narrowest sense. The opening sentences of her essay on Poe ('On Reading Poetry: Reflections on the Limits and Possibilities of Psychoanalytical Approaches', which appeared in her 1980 text, *The Literary Freud: Mechanisms of Defense and the Poetic Will*) pinpoint this approach. All of the questions of psychobiography we have encountered so far are put the other way around by Felman. Authors and readers do not make texts so much as texts 'create' their authors and readers. It is not with Poe as psychotic or neurotic that her concern lies, but with his poetry, of which mental aberration might itself be a symptom:

> To account for poetry in psychoanalytical terms has traditionally meant to analyze poetry as a symptom of a particular poet. I would here like to reverse this approach, and to analyze a particular poet as a symptom of poetry.[96]

Under discussion here is a wider question about how these discourses relate. '[T]he concept of "application"', Felman writes later in the Poe essay, 'implies a relation of *exteriority* between the applied science and the field it is supposed, unilaterally, to inform'.[97] But psychoanalysis is not so separate or so lofty; indeed, its own models bear this out, showing the implication of the desire and drives of reader, writer and text in any act of criticism, muddying the waters of ostensibly clearly directed applied readings. There can, then, be no such thing as an objective or transparently proven diagnostic reading, and it is psychoanalysis which offers the most complex models for how and what the critic (including, or perhaps especially, the naïve psychoanalytic critic) as well as writer brings to the text. In her important 1977 essay on Henry James' short story, *The Turn of the Screw*, Felman suggests that Freudian readings are anything but prescriptive or diagnostic:

> A "Freudian reading" is thus not a reading guaranteed by, grounded in, Freud's knowledge, but first and foremost *a reading of Freud's "knowledge,"* which as such can never *a priori* be assured of knowing anything, but must take its chances *as* a reading, necessarily and constitutively threatened by error.[98]

For psychoanalysis (as for Oscar Wilde), the truth of the text is seldom pure and never simple.

96 In *The Purloined Poe*, p. 133.
97 Ibid., p. 152.
98 'Turning the Screw of Interpretation', in *Literature and Psychoanalysis*, p. 116.

It is, then, psychoanalysis itself which offers a model for how the 'applying' psychoanalytic critic is already drawn deep into the terrain from which her 'science' would purport to be separate. In her short opening statement to the crucial 1977 collection, *Literature and Psychoanalysis: The Question of Reading: Otherwise*, Felman (who also edited the volume, which includes the James essay cited above) proposes the reinvention of the 'mutual relationship between literature and psychoanalysis'[99] in order to deconstruct the subordinative connection,

> in which literature is submitted to the authority, to the prestige of psychoanalysis. While literature is considered as a body of *language* - to be *interpreted* – psychoanalysis is considered as a body of *knowledge*, whose competence is called upon to *interpret*. Psychoanalysis, in other words, occupies the place of a subject, literature that of an object; the relation of interpretation is structured as a relation of master to slave . . .[100]

This is, of course, from the perspective of the ostensibly epistemologically 'superior' psychoanalysis, which lays claim to some form of scientific origin, and has a history of 'using' literary texts to bear out its theories. Felman, and other writers represented in the volume, proposes a reassessment of the relationship from the point of view of the literary, a fresh analysis of psychoanalysis as also a 'literary' practice:

> In the literary critic's perspective, literature is a subject, not an object; it is therefore not simply a body of language to interpret, nor is psychoanalysis simply a body of knowledge with which to interpret, since psychoanalysis itself is equally a body of language, and literature is also a body of knowledge . . .[101]

Felman thus proposes, and the essays in the volume play out, 'a real exchange, . . . a real dialogue between literature and psychoanalysis, as between two different bodies of language and between two different modes of knowledge'.[102] As part of a wider wave in cultural studies which has stressed the importance of interdisciplinary study as a site for interrogating the identities and differences of discourses, it is this debate which has had most impact on psychoanalytic criticism in the 1980s and 1990s.

These concerns are borne out in Felman's slightly later discussion of Poe. It is not surprising that Poe should become an object of specific interest to critics engaged in importing the work of contemporary French

99 Felman, 'To Open the Question', p. 5.
100 Ibid., p. 5
101 Ibid., p. 6.
102 Ibid.

theorists to America in the 1970s and 1980s, critics who are also working on those writers such as Mallarmé, Baudelaire and Valéry, who were themselves profoundly influenced by Poe from the mid-nineteenth century onwards. Felman and Johnson in particular have produced sophisticated readings of French poetry which engage with Derrida and Lacan in ways which have opened up the importance of both to English-speaking, primarily 'literary' audiences. Indeed, the poststructuralist marriage of deconstruction (a form of Continental philosophy) and Lacanianism (a form of psychoanalysis) has taken place largely across literary studies (which has traditionally thought of itself as separate from these disciplines). That Poe is one of the prime sites for this 'marriage' is deeply ironic, given the specific disagreement of Derrida and Lacan over Poe's work. Poe thus becomes a site not just for the collision of psychoanalysis and literature but for the interconnection of Continental philosophy and semiotics, French readings of Freud, American criticism and French poetry.

This interconnection mediated by Poe is traced by Barbara Johnson in 'The Frame of Reference: Poe, Lacan, Derrida'. Here, the nexus of these three writers produces a set of digressions and replications which place the reader in 'a vertiginously insecure position':[103] 'an unusually high degree of apparent digressiveness characterizes these texts, to the point of making the reader wonder whether there is really any true subject-matter there at all'.[104] Upon the very ambiguous foundation of the devious text itself is built a debate of charge and countercharge, focused on the presence, absence or identity of the story's 'truth'. Each of Johnson's texts – Derrida, Lacan and the Poe original – fends off the possibility of its own truth through deferral, orbiting or even disclaiming what we might expect analysis to focus on and reveal about a text – its final meaning, its core truth – by instead playfully and uneasily foregrounding the process of analysis itself:

> In all three texts, it is the act of analysis which seems to occupy the centre of the discursive stage, and the *act of analysis of the act of analysis* which in some way disrupts that centrality.[105]

Clearly this is a debate which promises to offer only frustration to the reader who wants the text to get to the point. Tracing through the game of cat and mouse in which Lacan and Derrida engage in their debate over Poe, Johnson identifies how these differences mark a significant interpretative gap, as well as a replication of the moves of the story itself:

103 Johnson, *The Critical Difference*, p. 110.
104 Ibid., p. 111.
105 Ibid., p. 110.

> The rivalry over something neither man will credit the other with possessing, the retrospective revision of the origins of both their resemblances and their differences, thus spirals backward and forward in an indeterminable pattern of cancellation and duplication.[106]

What this produces is, however, a possibility of reading which lies *between* these mismeetings, engendered precisely by the *failure to connect* which characterizes the debate (a mismeeting and failure to connect which is itself a repetition of the lost connections in the story itself). 'Failure' once again emerges as a key term to signal not an area with which the critic can do nothing, but fresh interpretative possibilities. Johnson's reading – indeed, her *way* of reading – has become so influential precisely because she opens up the possibilities implied by these mismeetings in such extraordinary ways: 'It is not how Lacan and Derrida meet each other' which is interesting here, 'but how they miss each other that opens up a space for interpretation'.[107]

Underpinning Derrida's critique of Lacan's essay on Poe is a much wider discussion about the relationship between deconstruction and psychoanalysis, a complex issue which rests upon Derrida's reading of Freudian and Lacanian psychoanalysis through a critique of logocentrism. Readers interested in pursuing this subject have a number of sources available to them now; rather than offer yet another survey, I shall sketch the few co-ordinates which will allow us to make most sense of Johnson's intervention between Lacan and Derrida.

As well as accusing Lacan of a general laxity in his deployment of philosophical sources,[108] Derrida was concerned to identify and deconstruct the key metaphysical pattern mapped out by classic Oedipal triangulation examined briefly in Chapter 1. The model family structure which traces the child's developing Oedipality is three-cornered, with the father intervening as the 'third term' to disrupt (in the child's eyes) the relationship of mother and child. But the triad might be read in other ways too: through his subsequent anxiety, the child sees himself as the intervening factor between father and mother (hence the father's role as threat), whilst it is the mother who represents the love-object for both father and child. In each pattern, a third person intervenes in the dyad of unity, introducing rivalry to the pattern.

106 Ibid., p. 118.
107 Ibid., p. 119.
108 See Derrida's *Positions* for an account of Lacan's 'light-handed reference to the authority of phonology, and more precisely to Saussurian linguistics', his 'philosophical facileness', and his 'art of evasion' in the face of 'theoretical difficulties'. Here Derrida also discusses the return to Freud as involving 'a massive recourse to Hegelian conceptuality', and his 'reinstallation of the "signifier," and psychoanalysis in general, in a new metaphysics ... in the space of what I then was delineating under the name of logocentrism' (*Positions* (London: Athlone, 1981), pp. 108–9).

On one level this triangulation might seem to present a topographical critique of binary structures – the 'either/or' dynamic upon which the logocentrism which Derrida deconstructs rests. The Oedipal structure is an interrelationship with three corners (or moments), not two. How then do we understand Derrida's charge, made in his essay on Poe ('The Purveyor of Truth'), that Lacan is operating with an implicit Hegelian brief, that psychoanalysis itself is founded upon a family structure which is itself at root logocentric? For Derrida, the 'third term' of the Oedipal triangle does not act to deconstruct the relationship between the other two positions (which would mean that the father's intervention unravelled the network of family relationships in which he is implicated). Rather, the Freudian (and Lacanian) 'third term' is incorporated in an Hegelian moment of synthesis, or *Aufhebung*. Of this Johnson writes, '[t]he problem with psychoanalytic triangularity, in Derrida's eyes, is not that it contains the wrong number of terms, but that it presupposes the possibility of a successful dialectical mediation and harmonious normalization, or *Aufhebung*, of desire'.[109] The 'third term' – the intervention of the father, threatening the mother–child dyad – offers not a breaking open but a continuation and consolidation of the family structure. The father comes in a spirit of disruption (disrupting a taboo desire), but acts out a moment of mediation between the two terms, moving the family form, unchanged, into the future and the next moment (for Hegel, the moment of synthesis or resolution of a dialectic forms the first 'thetic' moment of the next dialectical phase):

> The three terms in the Oedipal triad enter into an opposition whose resolution resembles the synthetic moment of a Hegelian dialectic. The process centres on the phallus as the locus of the question of sexual difference; when the observation of the mother's lack of a penis is joined with the father's threat of castration as the punishment for incest, the child passes from the alternative (thesis vs. antithesis; presence vs. absence of penis) to the synthesis (the phallus as a sign of the fact that the child can only enter into the circuit of desire by assuming castration as the phallus's simultaneous presence and absence; that is, by assuming the fact that both the subject and the object of desire will always be substitutes for something that was never really present).[110]

The child's desire thus acts as a totalizing moment of incorporation rather than a sign of difference. At the moment at which he learns of the substitution involved in signification, he consolidates a family structure articulated by Lacan – in Johnson's words – 'within the bounds of the type of "logocentrism" that has been the focus of Derrida's entire deconstructive enterprise'.[111]

109 *The Critical Difference*, p. 122.
110 Ibid.
111 Ibid.

That there are dialectical traces at work in Freud and Lacan is not exactly an original point. Freud debates this issue in some form in a number of places – for instance, his theory of instincts arguably rests upon a struggle with dualism which is not clearly resolved, particularly with the development of the theory of the death drive. The theory of the gaze as an aspect of the Other is also originally developed in Lacanian theory from the dialectical work of Jean-Paul Sartre. But – and this is Johnson's question – 'What does all this have to do with "The Purloined Letter"?'. There are two ways of answering this which have particular pertinence here. One of Johnson's arguments is that Derrida reads the first pattern of triangulation which Lacan identifies in the Poe story (its primal scene, involving the King, the Queen and the Minister) as both a straight translation of Freudian Oedipality and as only dialectically connected to the second triad, the resolving repetition (involving the Queen, the Minister and Dupin). 'The triangle', Johnson writes,

> becomes the magical, Oedipal figure that explains the functioning of human desire. The child's original imaginary dual unity with the mother is subverted by the law of the father as that which prohibits incest under threat of castration. The child has "simply" to "assume castration" as the necessity of substitution in the object of his desire (the object of desire becoming the locus of substitution and the focus of repetition), after which the child's desire becomes "normalized." Derrida's criticism of the "triangles" or "triads" in Lacan's reading of Poe is based on the assumption that Lacan's use of triangularity stems from this psychoanalytic myth.[112]

For Johnson, Derrida fails to see the repetition in Lacan's reading: these triads do not add up to one grand Oedipal narrative, but need to be read as interconnected but different moments. Lacan's final model is here quadrangular – the shape of two triangles placed together, with one a repetition (with a difference) of the other – but, Johnson argues, it is Derrida who is blind to this (a blindness brought about through what Johnson calls 'strategic necessity'[113] which needs to engage in a certain kind of critique). Derrida, in short, does not see the difference. How this matters to our reading of the text as an analytic work, analysing itself, acting on its readers, engendering debate and – for Lacan – a narrative of the flight of the signifier, lies in how we read the story as an allegory of the analytic situation.

It is in terms of the second of these issues that Johnson's title comes into play, with a discussion of the literary as a specially demarcated form of language. Derrida criticizes Lacan for failing to account for the 'literariness' of the Poe story – its narrative devices (the way in which it 'frames' itself), the context of other Poe stories – in order to read the story

112 Ibid., pp. 119–20.
113 Ibid., p. 125.

as a purely analytic narrative. One of Johnson's concerns is to bring both of these possibilities together – a narrative read for its analytic interest but precisely *as* fictive or poetic writing. I have said that the letter, for Lacan, is important not because of its content but because of its movement – its content is absent. Derrida, however, turns this absence into another kind of truth. In Johnson's words: 'In reading "The Purloined Letter" as an allegory of the signifier, Lacan, according to Derrida, has made the "signifier" into the story's truth.'[114] With this moment of consolidation, the text loses its indeterminate textuality, despite Lacan's emphasis on the mobility of elements within the story. In turning it into a narrative tracing the path of the signifier, even if this is a narrative which refuses to pin the signifier down, Lacan paradoxically gives significance to the lack of significance, and in this, if in nothing else, he deprives the text of its playful 'literariness'. But more important than this for Derrida is Lacan's lack of interest in the multiple literary references which frame, and are contained by, the text, the sense in which it is obsessed with the written word and transgresses the limits of the story itself, as it refers and spills beyond itself. For Derrida the story is a library, but it is Johnson who pulls 'some of the books off the shelves [to] see what they contain'.[115] For Johnson, then, Poe's text becomes an 'inter-text', as she reads its references and cross-influences. For Derrida, too, Lacan's text is an 'inter-text' – not entirely his own work. In the midst of several different critiques of psychobiographical 'decoding', comes a strange re-emergence of Marie Bonaparte's strategies; Bonaparte ironically becomes one of the crucial intertextual reference points between these texts:

> The very Oedipal reading that Derrida attributes to Lacan is itself, according to Derrida, a purloined letter – purloined by Lacan from Marie Bonaparte's psychobiographical study of the life and works of Edgar Allan Poe.[116]

In this apparently endless chain of intertextuality, are all texts then purloined versions of each other? Felman's reading of Poe has a different emphasis. She is not only concerned with the body of the text – or two bodies of texts (the literary and the psychoanalytic) – but with the effect of the text on the body of the reader, and with the reader's desire for the text. Poe has had a singular reading history, which Felman traces. Adored by an army of passionate readers (the French Symbolists included), the effect of his work on devotees has been like a poetic narcotic, as the fervent and unsettling psychic states, imagery and emotions act on the reader like a form of literary infection, transferred through the page and eye. He is the sublimely 'affecting' poet who "is experienced as

114 Ibid., p. 130.
115 Ibid., p. 132.
116 Ibid., p. 134.

irresistible" by his celebrants just as "in literary history, the poetry [is] most *resisted*".[117] At the same time, Poe's work has engendered equally vehement resistance, with early Freudian critics diagnosing the writer's 'sickness' through the work at the same time as the poetic is equated with the psychotic.[118] Haunting these encounters is the age-old yoking of poetry and madness, long theorized in literary discussions of creative states and now reworked in terms of a debate about knowledges, the blindness of one discourse to another. After engaging in her own strong resistance to Krutch's 1926 reading of Poe as exemplary literary neurotic, Felman pinpoints what is for her the key issue, which lies

> not so much in the interrogation of whether or not all artists are necessarily pathological, but of what it is that makes of *art* – not of the artist – an object of *desire* for the public; of what it is that makes for art's *effect*, for the compelling power of Poe's poetry over his readers. The question of what makes poetry lies, indeed, not so much in what it was that made Poe write, but in what it is that *makes us read him* and that ceaselessly drives so many people to *write about him*.[119]

Whilst this begins to articulate one point of connection between psychoanalytic critics and reader-response theory, the phenomenology of the reader's experience at the hands of the writer is not Felman's primary concern here. More important is a general question about compulsion, how our engagement as readers with a drive concerning, to absorb critically, to take in and read on, pushes forward understanding of the drive itself.

The problem with Krutch's reading is its importation of an external body of knowledge which is then applied to the text regardless of the interaction which takes place in the process of reading itself. It is thus an exemplary case of 'application', and as such is prone to all the problems that Felman is keen to identify. A theory of neurosis pre-dates the reading of Poe, making any subsequent reading inevitably cut to the measure of this prior 'knowledge': Krutch 'presupposes knowledge as a *given*, external to the literary object and imported into it, and not as a result of a reading-process, that is, of the critic's work upon and with the literary text'.[120] For Felman, both here and in the James essay, the importance of psychoanalysis is precisely that it unsettles fixed epistemologies which would impose a governing or finally determining frame of truth onto their subjects:

117 Felman in *The Purloined Poe*, p. 137.
118 Felman: 'Let us remember how many readers were unsettled by the humiliating and sometimes condescending psychoanalytic emphasis on Poe's "sickness," as well as by an explanation equating the poetic with the psychotic' (*The Purloined Poe*, p. 150).
119 Ibid., pp. 140–1.
120 Ibid., p. 140.

> Krutch . . . reduces not just Poe but analysis itself to an ideologically biased and psychologically opinionated caricature, missing totally (as is most often the case with "Freudian" critics) the *radicality* of Freud's psychoanalytic insights: their self-critical potential, their power to return upon themselves and to unseat the critic from any condescending, guaranteed, authoritative stance of truth.[121]

Felman's frequent concern, as we have seen, is with the interface of psychoanalysis and literature, not simply through connection and similarity but also through difference. A confusion of these terms thus characterizes Bonaparte's reading, which suffers from a 'blind non-differentiation or confusion of the poetic and the psychotic'[122] which has allowed the procedures of literature and psychoanalysis to blur into each other; Bonaparte is 'unaware of the fact that the differences [between psychoanalysis and literature] are as important and as significant for understanding the meeting ground as are the resemblances'.[123] Finally, Felman's problem with 'vulgar Freudianism' is its wilful need to trace 'poetry to a clinical reality', which constitutes a final explanation. The causal relationship which would link individual 'sickness' (or, for that matter, 'genius') to text in the fashion of explanation needs to be fundamentally challenged. So it is to the texts themselves that Felman turns for an account of the interface between psychoanalysis and literature which reduces neither one to the other, but which can nevertheless address the simple series of questions she poses:

> Is there . . . a way – a different way – in which psychoanalysis can help us to account for poetic genius? Is there an alternative to applied psychoanalysis? – an alternative that would be capable of touching, in a psychoanalytic manner, upon the very specificity of that which constitutes the poetic?[124]

The contrast is therefore between the immediacy of desire experienced by those in whom Poe's work strikes a passion and the distancing which critical history has set up, which makes him for Felman 'a unique literary case history', not simply because of these reading differences but because they are based in the direct 'effects' which reading Poe produces in the reader:

> Because of the very nature of its strong "effects," of the reading *acts* that it provokes, Poe's text (and not just Poe's biography or his personal neurosis) is clearly an analytical case in the history of literary criticism, a case that

121 Ibid.
122 Ibid., p. 142.
123 Ibid.
124 Ibid., p. 144.

> suggests something crucial to understand in psychoanalytic terms. It is therefore not surprising that Poe, more than any other poet, has been repeatedly singled out for psychoanalytical research, has persistently attracted the attention of psychoanalytic critics.[125]

When Felman turns to Lacan it is for this reason – his insistence on the effect of the letter in Poe's story rather than anything else. In contrasting Lacan's work on Poe with Marie Bonaparte's, Felman is able to stress again that what matters is the impact of the text on subjects rather than the analysis of words-as-symptoms of a writer's psychic state, in a way which 'radically subverts not just th[e] clinical status of the poet, but along with it the "bedside" security of the interpreter'.[126] The clearly demarcated relationship of each to the other – poet and critic as patient and analyst – is thus under scrutiny. If Bonaparte's project is to search for the *content* of the purloined letter as prime symptom of Poe's buried self ('*hidden* somewhere in the real' as Felman puts it, 'in some secret biographical *depth*'),[127] Lacan's is the opposite: to trace the effect movement of and in the text has on the interconnected subjects of the tale. 'The history of reading', Felman writes,

> has accustomed us to the assumption . . . that reading is finding meaning. . . . Lacan's analysis of the signifier opens up a radically new assumption, an assumption that is nonetheless nothing but an insightful logical and methodological consequence of Freud's discovery: that what *can* be read (and perhaps what *should* be read) is not just meaning, but the lack of meaning; that significance lies not just in consciousness, but, specifically, in its disruption; that the signifier can be analyzed in its effects without its signified being known; that the lack of meaning – the discontinuity in conscious understanding – can and should be interpreted as such, without necessarily being transformed into meaning. . . . Thus, for Lacan, what is analytical par excellence is not (as is the case for Bonaparte) the *readable*, but the *unreadable*, and the *effects* of the unreadable. What calls for analysis is the insistence of the unreadable in the text.[128]

This ability to find, if not meaning, then interpretable material, in meaninglessness certainly comes close to confirming Derrida's argument against Lacan, which is that at the heart of this whole process a residual, if negatively expressed, need for meaning remains. If what can or should be read 'is not just meaning, but the lack of meaning', analysis can only shift between these two poles in what is essentially a dialectical choice, finding something fixed – a presence or an absence of meaning – with

125 Ibid., p. 137.
126 Ibid., p. 150.
127 Ibid., p. 148.
128 Ibid., p. 149.

either possibility. But Felman's twists on the issue makes the important term here not meaning (or lack of it) but *interpretation*. Thus texts which lack any overtly meaningful content – the letter, for instance – have as much interpretability as those which seem pregnant with significance, and their 'lack' becomes not a negative term but a moment of interpretable discontinuity which challenges our desire to *know*. Here again the James essay has a purchase, since it is here that Felman most clearly stresses the importance of reading with 'tact', against the 'vulgarity' of a literal reading. Taking the terms of James's own narrative deferrals as her cue, she develops a reading which resists drawing fixed truths from the text, instead allowing the text to withold its secrets, continuing to circulate its ambiguities and discontinuities. At the communal story-telling session with which *The Turn of the Screw* begins, Douglas, the narrator, responds to demands for a story which offers its meanings as crude and blunt revelations with a sharp statement of Jamesian poetic practice:

> Mrs Griffin, however, expressed the need for a little more light. 'Who was it she was in love with?'
> 'The story will tell,' I took upon myself to reply.
> 'Oh, I can't wait for the story!'
> 'The story *won't* tell,' said Douglas; 'not in any literal, vulgar way.'
> 'More's the pity, then. That 's the only way I ever understand'.[129]

The implication is clear (even if little else is in this story): if you are one such reader, who only understands the literal and the vulgar, you would do better to stop reading now – this is a story which will inevitably defy you. 'Vulgarity' as a reading practice is then established as deeply anti-Jamesian, with deferral, ambiguity, 'not-telling', as the preferred terms of interpretation, inside the text and out, for Felman takes her reading tools from the story's own operations of 'tact'. Felman's role here is thus to return to the text's own way of reading itself, in order to reread what is at stake in the history of vulgar Freudian readings which have precisely ignored James's advice, trying to pin down suggestions into diagnostic 'tellings'. James, and Felman's alternative psychoanalytic critical practice, would restore the silences to the text so as to allow its movements once again. Against vulgar, literal readings she writes:

> The literal is the "vulgar" because it *stops* the *movement* constitutive of meaning, because it blocks and interrupts the endless process of metaphorical substitution. The vulgar, therefore, is anything which misses, or falls short of, the dimension of the symbolic, anything which rules out, or excludes, meaning as a loss and as a flight, – anything which strives, in other words, to eliminate from language its inherent silence, anything which misses the specific way in

129 Henry James, *The Turn of the Screw* (Harmondsworth: Penguin, 1982), pp. 9–10.

> which a text *actively* "won't tell." The vulgarity that James then seeks above all to avoid is that of a language whose discourse is outspoken and forthright and whose reserves of silence have been cut, that of a text inherently *incapable* of silence, inherently unable to hold its tongue.[130]

It is clear from this why Lacan's diffuse, mobile reading of Poe should be so seductive to a critic working with the sense of 'literariness' as a possibility of active un-saying, suggestion, impossibilities in language, which are symbolically telling but which nevertheless resist a ready process of decoding. However, this is clearly not the final word on the interconnection of psychoanalysis and literature or film. Felman's statement on their 'interimplication' in each other at the end of her essay on Poe is clearly important, but this is only part of the frame I wish to establish for the readings below in Chapters 3–6. Krutch, Bonaparte, Kuttner on Lawrence, Freud on Leonardo – the list of critical 'vulgarity' is inordinately long, and these texts remain important, despite (or perhaps because of) the stark determinism with which they make their case. Against this, Felman, reading Lacan on Poe, identifies a breakdown of boundaries taking place:

> [S]ince psychoanalytic theory and the literary text mutually inform – and displace – each other; since the very position of the interpreter – of the analyst – turns out to be not *outside*, but *inside* the text, there is no longer a clear-cut opposition or a well-defined border between literature and psychoanalysis: psychoanalysis could be intraliterary just as much as literature is intrapsychoanalytic. The methodological stake is no longer that of the *application* of psychoanalysis to literature, but rather, of their *interimplication in* each other.[131]

Underpinning this is a question about what we want from texts, how our need for mastery over the text is facilitated, or resisted, by the ease or difficulty with which the text allows 'entry'. '[I]f we found reading hard it was partly because of the strength of our own latent wishes,' writes Daniel Gunn in his analysis of Serge Leclaire's analytic work, *A Child Is Being Killed*.[132] Here he outlines the steps taken in an exemplary psychoanalytic reading. Such a reading is one which operates on the reader, and the writer, as does the (Lacanian) analytic session: disrupting a sense of mastery, control and centredness. This form of reading psychoanalytically thus takes place through the difficult process of transference and countertransference, which Leclaire's text, like many an avant-garde counterpart, enables. The book is obscure, it denies its reader immediate mastery, it witholds its referents, its meanings are internal, its sense slides. '[I]t is indeed impossible to place, to know, or to understand,

130 Felman, 'Turning the Screw of Interpretation', p. 107.
131 In *The Purloined Poe*, pp. 152–3.
132 *Psychoanalysis and Fiction* (Cambridge: Cambridge University Press, 1990), p. 42.

because the co-ordinates are withheld which might allow us to place from where this text is being spoken (or written).' Replaying the findings of Freud's original essay of 1919, 'A Child Is Being Beaten' (which essentially investigates the origin of this strange statement, asking who is speaking it to whom, before setting forth a model of the subject fragmented by fantasy), what emerges as the problem is our readerly desire for centred sense. Textual difficulty is, then, a way of generating a confrontation with one's own desires and expectations, even if that is because those desires are actively thwarted by the text: 'Leclaire guides his reader in an understanding of his or her reaction to the text', disarming us, as Gunn puts it, 'by revealing us as armed'.[133] What is being revealed in our frustrated confrontation with the obscure, witholding text, is the murderous critical desire *with which we operate* upon a text. It seems that reading can be murder.

This, indeed, is one of Felman's conclusions: the relationship between psychoanalysis and literature has often been viewed in murderous terms. When Felman charges against 'vulgar Freudianism' which would strip a text of its symbols, she is identifying an interpretative strategy marked by aggression. '[T]he propagation of psychoanalysis', writes Julia Kristeva, 'has shown us, ever since Freud, that interpretation necessarily represents appropriation, and thus an act of desire and murder.'[134] In her analysis of this problem, Jane Gallop also discusses how reading Lacan is an uncomfortable process which confounds the reader's desire to be the 'subject presumed to know' in relation to the text: '[I]nterpretation is always motivated by desire and aggression, by desire to have and kill.'[135] If she cannot finally 'know her 'text', the reader cannot claim a superior or privileged position of authority over it; she cannot murder to dissect. Again, the vulgar Freudian symbolist is demonized, this time by Gallop: 'The "symbol hunter" identifies with the position of knowledge, takes on the illusory role of "someone who knows," who knows the unconscious, who knows what the other, the author, does not know'.[136] Against this is posited a textuality precisely engaged in *resisting* fixed analysis.

In the remainder of this book I shall open up a number of texts – novels, films, poems, and an autobiography – in ways which will be generally resistant to mastery (although they might sometimes be a little bit 'vulgar'). Alongside the strictly literary I will read some key psychoanalytic texts too, with the aim of encouraging the kind of textual cross-fertilization which seems to me to get the most from both sets of texts, as well as to explore further what seems to be a new avant-garde field of

133 Ibid., p. 41.
134 Julia Kristeva quoted by Gallop, *Reading Lacan*, p. 27.
135 *Reading Lacan*, p. 27.
136 Ibid., p. 29.

writing, the psychoanalytic essay or lecture, itself mimicking the analytic situation in its replay of patterns of transference and their accompanying difficulties. For Felman, 'Poetry . . . is precisely the effect of a deadly struggle between consciousness and the unconscious';[137] I want to explore now whether similar struggles also take place over other forms of writing.

137 In *The Purloined Poe*, p. 154.

3

Writing at Play: Fantasy and Identity in Angela Carter

Angela Carter has been hailed as the writer of literature's unconscious. Her fictions, in Lorna Sage's words, 'prowl around on the fringes of the proper English novel like dream monsters – nasty, erotic, brilliant creatures that feed off cultural crisis'.[1] 'Seeing is believing' is the challenging refrain of Carter's 1986 work, *Nights at the Circus*, a novel which overtly explores the possibility that subjective identity is constructed not in the real world of empirically realized 'fact' but through fantasies of self and sex. Identity – the identities of characters, of narrative voices, even of the readers themselves – is a protracted experiment in the artful deployment of suspension of disbelief: it's all a matter of how successfully, and for how long, the self can play out its masquerade of persona. The construction of the subject and its sex is centre-stage in Carter's work, the prime spectacle in her circus-ring. Fevvers, the heroine of *Nights at the Circus*, is a woman with wings, or so those around her believe, since this is what they (and we) see – their senses have tricked them into believing the impossible. If psychoanalysis believes in fantasy, defying the primacy of empirically verifiable truth, Carter subverts the whole show by offering an empirically verifiable truth (a woman with wings) which can only be the stuff of fantasy. For Juliet Mitchell, 'Freud started believing that the stories were real, and then believed that they were real as stories'. *Nights at the Circus* enacts a sleight of hand which turns the real world *into* a story. Carter's dream-monsters invade the space governed by empiricism, saying,

1 Lorna Sage, 'A Savage Sideshow: A Profile of Angela Carter', *New Review* 4/39–40 (July 1977), p. 51.

> Look at me! With a grand, proud, ironic grace, she exhibited herself before the audience as if she were a marvellous present too good to be played with. Look, not touch.
>
> She was twice as large as life and as succinctly finite as any object that is intended to be seen, not handled. Look! Hands off![2]

Thus Carter makes the real world subject to the order of fantasy. Playing with the psychoanalytically questionable distinction between the real world of sense and the fantasy world of dreams, she collapses one into the other. Fevvers's fantastic world, unreasonable as it is, is built upon the empirical assumption that what you see is what you get – what you see *is* what is – a sensory gullibility which is ripe for subversion on its own terms. If the wound is healed between appearance and reality, anything is possible, anything might become real, as long as it is a thing which can 'appear'. This unsettling terrain is the site of Carter's work, and it is here that two ways of understanding fantasy connect.

Fantasy is one realm within which cultural categories and psychoanalytic formulations explicitly collide and collaborate, for fantasy – as genre and as unconscious state and process – is a term which has a clear purchase in both areas. In its latter form, fantasy is the bedrock of psychoanalysis. It is through a stock of constantly revised fantasy-narratives, scenarios and memories that the self comes into being. If fantasy is crucial to the psychoanalysis of the subject, it is also fundamental to writing, cultural representation and imagination, and cinema. We already have some sense of its psychoanalytic importance from the above discussion of Freud's emphasis on primal scenes, the role of fantasy in the sexualization of the infant, as well as dream interpretation. Fantasy is, however, also a literary and visual genre, and Carter's work intersects both realms. Whilst some of her texts could generically be categorized as fantasy-works, her writing also engages with discussions of the sexualized subject coming directly from, and pertaining to, psychoanalytic theory. Carter is a fantasist in more ways than one; her dream-monsters are perfectly placed to demonstrate the relationship between two types of fantasy, both of which rely upon visions of the impossible; in Stephen King's words, she fulfils 'the primary duty of literature – to tell us the truth about ourselves by telling us lies about people who never existed'.[3] These are, however, very slippery 'truths', and Carter wants us to challenge as well as believe in the lies. This chapter will explore the connection between Carter's literary fantasy of a world turned upside down, in which seeing is believing, and the underworld of unconscious fantasy based upon a bedrock of primal scenes. I will look at Carter through the 'impossible' visions of Freud's fetishist and the primal fantasies of his

2 *Nights at the Circus* (London: Chatto & Windus, 1984), p. 15.
3 *Danse Macabre* (London: Macdonald, 1992), p. 282.

paranoid woman (read partly in a 'pornographic' connection with Anaïs Nin). Later in the chapter I do, however, want to turn to a very different tradition of cultural analysis, coming from Melanie Klein and object-relations theory. As well as opening up some further connections between culture and play, this will also lead into some discussion of alternative feminist psychoanalytic visions of impossible or utopian spaces, deriving from a return to the pre-Oedipal connection with the mother, as well as from the work of Nancy Chodorow.

I
Seeing Is Believing: *Nights at the Circus*

Angela Carter uses the possibilities of fantasy to challenge the difference between appearance and reality. To borrow from Rosemary Jackson's discussion of it as a literary genre, Carter is one manifestation of 'the inside, or underside, of realism'; in her work it is 'as if the novel had given rise to its own opposition, its unrecognizable reflection'.[4] Or perhaps not unrecognizable: like Lewis Carroll's Alice when she slips through the 'gauze' or 'mist' of the mirror into Looking Glass House, the shape of the old right-side-up world is still there, but in opposition. As Eric Rabkin writes in *The Fantastic in Literature,* 'The Truly Fantastic occurs when the ground rules of a narrative are forced to make a 180 degree reversal, when prevailing perspectives are directly contradicted.'[5]

Literary fantasy is thus predicated upon reversal, playing with the real and the reasonable to challenge expectations. The psychoanalytic understanding of the subject does, however, rest upon what is often understood as a very different type of fantasy rooted in the origins of psychic life. As we have seen, the self comes into being in a torrent of storytelling, as it repeats, creates, elaborates the myth of its own origin through primal scenes. We narrate ourselves into existence as subjects, and continue explaining and rewriting that existence through fantasy scenarios which can be conscious, unconscious, or somewhere in between. In 'Creative Writing and Day-Dreaming' (1908) Freud explicitly connects different types of fantasy indulged in at different phases of development. 'The creative writer does the same as the child at play',[6] Freud states, also relating the play of unconscious fantasy and 'waking reverie' (day-dreaming) with child's play and writing, so that all adult fantasy is read as a version of childhood wish-fulfilment acted out in

4 *Fantasy: The Literature of Subversion* (London and New York: Methuen, 1981), p. 25.
5 Quoted ibid., p. 21.
6 'Creative Writers and Day-dreaming' (1908 [1907]), PF vol. xiv, p. 132.

play ('The motive forces of phantasies are unsatisfied wishes, and every single phantasy is the fulfilment of a wish, a correction of unsatisfying reality.'[7]) '[W]e can never give anything up; we only exchange one thing for another,' he writes, drawing the connection more clearly:

> What appears to be a renunciation is really the formation of a substitute or surrogate. In the same way, the growing child, when he stops playing, gives up nothing but the link with real objects; instead of playing, he now phantasies. He builds castles in the air and creates what are called day-dreams. I believe that most people construct phantasies at times in their lives.[8]

Thus the connection between the unconscious fantasies of, say, Dora in the case of that name, and the creative act of writing it up, is made.[9] If Freud's case histories are stories, then dreams are stories too, and writing them is no less a 'creative' act. If Dora the dreamer is 'creating' a story in her fantasy, so Freud the writer is fantasizing in the very act of narrative construction.

An explicit connection between 'conscious' writing and 'unconscious' fantasy is thus already made in psychoanalysis itself. But what of writing also characterized by its fantastic content – its location in a genre? *Nights at the Circus* is one such text. It traces the picaresque path of its winged Cockney heroine from London to Moscow to Siberia, in the steps of the journalist Walser, who starts as the text's 'voice of reason', joins the circus to track ever more closely our heroine's progression, and plummets into his own hallucinatory psychic subversion at the hands of the Siberian Shaman as the novel draws to a close. As Walser's sense of control slowly unravels, so the story of Fevvers's strange origins and childhood is unwound, and she too moves towards a climactic moment of psychological uncertainty. Fevvers is playfully sketched as the crude dream-symbol of sexual pleasure, figuring as an ironic image for impending emancipation (which Carter knows full well never entirely comes): 'I was possessed by the idea that I had been feathered out for some special fate,'[10] she says, and Lizzie comments that she 'must be the pure child of the century that just now is waiting in the wings, the New Age in which no women will be bound down to the ground'.[11] If Fevvers

7 Ibid., p. 134. Because Freud explicitly draws this connection here between conscious and unconscious fantasy, I will use the term 'fantasy' (rather than 'phantasy') throughout this discussion, except where it appears with a 'ph' in quotations, since this discussion hinges on the way in which texts undo these differences. Readers should be aware that despite this some writers (particularly Kleinians) prefer to reserve 'phantasy' for its unconscious form.

8 Ibid., p. 133.

9 See Chapter 1 for a fuller discussion of the 'Dora' case as literature, especially the account of Steven Marcus's analysis.

10 Carter, *Nights*, p. 39.

11 Ibid., p. 25.

is partly the simple symbol of *jouissance* (her 'body was the abode of limitless freedom', fleshing out the common idea that dreams of flying are dreams of sex), she might also be the wish-fulfilment of another, particularly female dream-desire which Freud discusses, '"to be like a bird"; while other dreamers became angels during the night because they had not been called angels during the day'.[12] As hero and heroine, rational Western man and unreasonable feminine freak alike, confront the possibility that they may be nothing more than a fiction, so they also fall in love. This is all laced with a *fin de siècle* sense of possibility and impending apocalypse – the twentieth century begins as the novel ends.

Walser is then our rational reader-in-the-text,[13] our surrogate and representative (that is, if we are looking for logical explanations), the realist decoder and figure of reason encountering and desperately trying to resolve the mysteries of monsters. He fails spectacularly, and plummets into the 'unreasonable'. But like many of the earlier texts of modernism which *Nights* is parodying, this is a novel which initially seduces you into being caught off your guard. It races along, all the time desperately trying to arm you with arguments (hopefully provided by Walser) which might explain the visual enigma of Fevvers.

> LOOK AT ME!
> She rose up on tiptoe and slowly twirled round, giving the spectators a comprehensive view of her back: seeing is believing. Then she spread out her superb, heavy arms in a backwards gesture of benediction and, as she did so, her wings spread, too, a polychromatic unfolding fully six feet across.[14]

What is the audience to do with this? Fevvers' slogan 'Is she fact or is she fiction?'[15] is answered in a number of ways. Fraud or freak, or more charitably, mimic or secular miracle, she is a one-woman performance of the truth of the purloined letter, paradoxically putting herself on show in order to mask her monstrosity. As Walser reflects, 'in a secular age, an authentic miracle must purport to be a hoax, in order to gain credit in the world'[16] (this is a bit like the theory that Elvis is alive and hiding as an Elvis impersonator). Here Walser has tried to rationalize the image, but his account is never substantiated. Rather, the explanation hovers over

12 Freud, *The Interpretation of Dreams*, PF vol. iv, p. 518.
13 In his important analysis of *Stagecoach*, Nick Browne proposes the position of a 'spectator-in-the-text' who functions as a focus for our identification, as the on-screen seer of events in the film's diegesis. Walser similarly 'reads' and interprets the events of *Nights*, the reader's proxy in the narrative. See Nick Browne, 'The Spectator-in-the-Text: The Rhetoric of *Stagecoach*', in *Movies and Methods*, ii, ed. Bill Nichols (Berkeley, CA, and London: University of California Press, 1985), pp. 458–75.
14 p. 15.
15 p. 7.
16 p. 17.

Fevvers's story as a canny economic possibility: In order to be no one's commodity but her own, the winged woman must *not* opt to *conceal* her deformity. Rather she must *make a show of it*, demonstrating it as an *undecidable* thing, making a spectacle of herself as neither monster nor trickster but as something which might just be either. Taking an alternative path from those who live in the freak show Fevvers also briefly inhabits ('Amongst the monsters, I am well hidden; who looks for a leaf in the forest?'[17]), she places herself centre-stage but – by doing so – makes the reality of herself dubious. She hides behind the question 'Is she fact or is she fiction?' rather than hiding behind the crowds of others who are like her. Ambiguity will be her economic salvation – the bird-woman can continue to work, escaping the more negative options of categorization as medical aberration or exposure as fake.

Walser's rationalization does, however, have exactly the opposite effect to its intention. Rather than explaining away the bird-woman's ambiguous status, he highlights its indeterminacy, and underlines the sense in which this core element of *Nights at the Circus* epitomizes literary fantasy. In Tzvetan Todorov's important structuralist study of *The Fantastic,* it is argued that the central moment and experience of fantasy is *hesitation*. As Jackson puts it, summing up Todorov, 'the purely fantastic text establishes absolute hesitation in protagonist and reader: they can neither come to terms with the unfamiliar events described, nor dismiss them as supernatural phenomena'[18] – they hesitate in the gap in between two explanations, having to believe what they see without being able to explain it. When confronted with the spectacle of a bird-woman, Fevvers's audience *ought* to take one of two paths – in the words of Todorov:

> [E]ither he is the victim of an illusion of the senses, of a product of the imagination – and laws of the world then remain what they are; or else the event has taken place, it is an integral part of reality – but then this reality is controlled by laws unknown to us.[19]

'It is', as James Donald writes, 'the moment of uncertainty between these two possibilities that constitutes the fantastic. To opt for either possible answer is to slip from the fantastic into a neighbouring genre'[20] – the uncanny being one 'neighbouring genre' (when the disquieting thing can be finally explained rationally, as with Freud's uncanny), the marvellous being the other (when it is truly realized as supernatural). Fevvers

17 p. 65.
18 *Fantasy*, p. 27.
19 Tzvetan Todorov, *The Fantastic: A Structural Approach to a Literary Genre*, trans. Richard Howard (Ithica, NY: Cornell University Press, 1973), p. 25.
20 See James Donald, Introduction, *Fantasy and the Cinema*, ed. Donald (London: BFI, 1989), p. 11.

refuses to shake down into either one or the other, and so she continues to walk the tightrope of doubt which constitutes the fantastic, and we – alongside Walser – continue to 'hesitate between a natural and a supernatural explanation'. We cannot decide between the options of illusion (the uncanny) and supernature (the marvellous). We have to see the impossible – we have to see both and neither possibilities at the same time. The text encourages and keeps open a perpetual double reading, splitting the reader between two options which she must simultaneously – and pleasurably – maintain. The effect of fantasy in producing this splitting or doubling of response is thus not unlike that of the ego discussed by Freud in his short (and unfinished) but very important late paper, 'Splitting of the Ego in the Process of Defence' (1940 [1938]) (the starting-point for much of Lacanianism). Here the self is not split topographically – between, say, the ego, the id, and the super-ego as separate subjective domains – but rather the ego itself is read as capable of contradictory response ('the whole process seems so strange to us', Freud writes, 'because we take for granted the synthetic nature of the processes of the ego.' The ego is, however, 'liable to a whole number of disturbances'[21]). When faced with two contradictory demands between which he has to choose (an instinctual demand and the prohibition of that demand by the reality principle), the child

> takes neither course, or rather he takes both simultaneously, which comes to the same thing. He replies to the conflict with two contrary reactions, both of which are valid and effective.[22]

Whilst Freud is addressing specific circumstances here (one of which might be the threat of castration and the child's disavowal of this, which I shall discuss presently in relation to fetishism), this sense of a conscious awareness of contradictory impulses has a wider currency, but particularly in cultural analysis of texts which frequently require a theory which can account for 'a whole number of disturbances'. Texts are not unitary in their meanings and significances, particularly texts which open up the 'hesitations' of identity above all else. Neither are egos – nor, indeed, the responses of readers.

Let us keep in mind this possibility of 'taking both courses simultaneously' as we move further into the text, alongside Freud. It is this space of hesitation into which Walser sinks ever faster, especially once he meets the Shaman:

> (The Shaman) made no categorical distinction between seeing and believing. It could be said that, for all the peoples of this region, there existed no differ-

21 Freud, 'Splitting of the Ego in the Process of Defence', PF vol. xi, p. 462.
22 Ibid., p. 461.

> ence between fact and fiction; instead, a sort of magic realism. Strange fate for a journalist, to find himself in a place where no facts, as such, existed![23]

As he submits to the Shaman's mode of visual gullibility, the impossible thing that Walser sees is exactly what he has to believe. If Fevvers is the visual enigma, Walser is the fantastic subject: suspended in a moment of hesitation between resolving a thing as wild but finally explicable, or marvellously supernatural, he cannot consolidate an explanation: he must just submit to what he sees. What he learns in the fantastic, then, is how to 'hesitate' for long enough to believe what he sees, even when what he sees is an impossibility.

II
Fetishism and Visual Impossiblity

Freud has also written of men who see the impossible. 'Probably', he writes (with the tone of one about to make a claim for the best lager in the world) 'no male human being is spared the fright of castration at the sight of a female genital.'[24] Whilst the universal truth of this may itself be problematic, the model of male response to female absence has been enormously influential. Seeing is believing, and seeing that thing which isn't a thing must mean believing in the horror of the reality of castration. The sight of the penis-less woman signifies castration which could be the boy's own fate ('if a woman has been castrated, then his own possession of a penis [likened by Freud to his 'Throne and Altar'] was in danger'). Too disturbing to be believable, even if the vision tells you it's true, what the fetishist must do is rewrite the vision and see something else instead. So instead of believing that the thing (the penis) is not there, the boy inserts something else in its place. 'To put it more plainly,' writes Freud again, 'the fetish is a substitute for the woman's (the mother's) penis that the little boy once believed in and – for reasons familiar to us – does not want to give up.'[25] In short, the little boy who is to become the adult fetishist first sees the female genitals and cannot believe what he sees – the vision is loaded with the trauma of the possibility of castration. The fetish then stands in for the penis which isn't there, and the adult fetishist relates sexually to women in terms of the fetish-object. But the choice of object itself reveals something of the narrative which surrounds the original moment of shock: the fetish which comes to stand in for the woman's absent penis will be the last thing which the boy saw *before* the awful

23 *Nights*, p. 260.
24 Freud, 'Fetishism', PF vol. vii, p. 354.
25 Ibid., p. 352.

vision – the fetish-object is then a representation of the story's cut-off point:

> when the fetish is instituted some process occurs which reminds one of the stopping of memory in traumatic amnesia. As in this latter case, the subject's interest comes to a halt half-way, as it were; it is as though the last impression before the uncanny and traumatic one is retained as a fetish.[26]

Freud has shown an interest in these narrative cut-off points from his earliest work. In *Studies on Hysteria*, discussing 'the phenomenon of hysterical symptoms joining in the conversation' and interrupting the analysis, he raises a question of mental censorship and negation which has the effect of the suspense strategy of thriller fiction:

> Every newspaper reader suffers from the same drawback in reading the daily instalment of his serial story, when, immediately after the heroine's decisive speech or after the shot has rung out, he comes upon the words: 'To be continued.' In our own case the topic that has been raised but not dealt with, the symptom that has become temporarily intensified and has not yet been explained, persists in the patient's mind and may perhaps be more troublesome to him than it has otherwise been.[27]

But the fetishist's story is not 'continued', except with reference to an object which stands in for the final scene evidencing castration – this is a story whose ending will always be displacement rather than denouement. Freud goes on to catalogue likely fetish-objects, based on their location on the body as representative of the last moment before the 'absence' was discovered:

> Thus the foot or shoe owes its preference as a fetish . . . to the circumstance that the inquisitive boy peered at the woman's genitals from below, from her legs up; fur and velvet . . . are a fixation of the sight of the pubic hair, which should have been followed by the longed-for sight of the female member; pieces of underclothing, which are so often chosen as a fetish, crystallize the moment of undressing, the last moment in which the woman could still be regarded as phallic.[28]

By inserting an object into the gap, the little boy turns an impossibility into a sexual fiction. He puts a fetish in place of the absence, and gets off on that instead. Believe the thing that you see, or rather that you *now* see, now that you've put it there, rather than the meaning (castration) of the thing that you *didn't* see.

26 Ibid., p. 354.
27 Freud and Breuer, *Studies on Hysteria*, pp. 384–5.
28 'Fetishism', pp. 354–5.

The tale of the anxious boy who disavows 'his perception of the woman's lack of a penis'[29] and then continues to do so throughout his adult sexual life with the 'substitute sexuality' of fetishism has a wider importance. Freud's fetishistic little boy sees a thing which he cannot believe in, or at least cannot assimilate – fetishism is the 'very energetic action [which] has been undertaken to maintain the disavowal' of woman's 'lack'. The boy 'has retained that belief, but he has also given it up'. The penis is not there, something else is there: in the crisis moment he sees an absence and a presence at the same time. Crucially, the essay claims a universality for this experience ('Probably no male human being is spared . . .'), and thus Freud's account of how the boy copes with that inability to believe what he sees must then also be generally (or probably) true – few men, it seems, can see from the perspective of Carter's magic realist, who believes *untraumatically* in everything he sees. The Shaman lives outside this Freudian world of incredulous vision, and his untraumatic apprehension of the truth of all that he sees is symptomatic of the psychic utopia he inhabits. The Freudian boy's vision is a key unconscious fantasy, tainted by a distressing awareness of the violent significance of woman's 'lack'. The fetishist's moment of hesitation, where his visual disavowal takes place, is a specific response to castration anxiety, 'normally' avoided through a traumatic recognition that the woman's lack signifies the father's castrating power.

Upon this small story of visual shock rests a much bigger story about the construction of sexual difference. This scene in which the woman (who has, in actual fact, got *female* sexual organs, not an *absence* of the penis) is misrecognized and perceived as castrated man, as the bleeding image of the fact that castration is possible, occupies a crucial place in the psychoanalytic account of the development of sexual difference. The fetishist may be a specific case (Freud says that not all men cope quite so badly with the sight of a woman, and the normal male path is through acceptance of the possibility of castration), but the *vision* of difference is universally galvanizing. The woman is castrated, the boy has to cope with what this means, and so begins a process of identification and denial which culminates (he hopes) in the achieved power of full masculinity. I will return to this again shortly: upon a visual experience (or fantasized visual scene) rests the construction of identity.

What both of these texts have in common is that the unbelievable vision is of a woman. Whether it is a woman with wings or one with no penis, for the formative Western male subject (Walser or the Freudian boy) it is the woman who is visually unbelievable. And she is a fantasy – Fevvers of the literary kind, and the boy's castrated woman of the 'primal phantasy' variety. Carter is, of course, engaging in some sort of critique of the Walser mentality which sees woman as a problem on the

29 Ibid., p. 352.

evidence of her visual appearance. This is not to say that Carter's feminism enacts a traditional critique of the male gaze and its objectification of women – on the contrary, Carter loves the visual world, and her heroines frequently engage in a wild celebration of their looks. When asked by Lorna Sage in 1977 if her 'people's appearances are their essence', she replied, 'Of course, it's a world of appearances. I call this materialism. . . . Everyone's appearance is their symbolic autobiography.'[30] As a feminist negotiation of the dynamics of vision and the construction of identity, Carter's work suggests a fantasy of feminine identity, 'made up' from a woman's point of view. In a splendid example of feminist false consciousness, Fevvers exploits her image to the hilt, writing this 'symbolic autobiography' with the repeated phrase 'Look at me!' (positively *inviting*, then, the glimpse of the all-seeing tin can in Lacan's paranoid anecdote[31]).The problem, then, isn't *that* the other looks but *how* he looks. There is a moment near the end of the novel when Fevvers almost becomes the victim of this:

> In Walser's eyes, she saw herself, at last, swimming into definition, like the image on photographic paper; but, instead of Fevvers, she saw two perfect miniatures of a dream.
>
> She felt her outlines waver; she felt herself trapped forever in the reflection in Walser's eyes. For one moment, just for one moment, Fevvers suffered the worst crisis of her life: 'Am I fact? Or am I fiction? Am I what I know I am? Or am I what he thinks I am?'[32]

But as a true Carter heroine, Fevvers is able to assert her bizarre visual reality in a sure move which overcomes this threatened objectification. Other women sometimes succumb to an objectifying male gaze in Carter; the heroine of 'The Bloody Chamber', for instance, is chosen by her Bluebeard husband for her likeness to a sado-masochistic etching he owns. She is repeatedly confronted by images of her own flesh in his eyes, seeing herself 'suddenly, as he saw me, my pale face, the way the muscles in my neck stuck out like thin wire'. Even when he is gone she is reminded of his surveillance as he haunts the mirrors in which he has seen and caught her (her bedroom 'retained the memory of his presence trapped in the fathomless silvering of his mirrors'[33]). She escapes this only when rescued by her mother (and the love of a blind man). Fevvers escapes her moment of doubt by flapping her wings, but not before she has confronted her own impossibility, with Walser transformed by

30 Sage, 'A Savage Sideshow', pp. 55–6.
31 See discussion of this in Chapter 2 above, p. 73.
32 *Nights*, p. 290.
33 Carter, 'The Bloody Chamber', in *The Bloody Chamber and Other Stories* (1979) (Harmondsworth: Penguin, 1985), p. 30.

Fevvers' fear into another of Carter's Bluebeards who would fix the woman to *his perception of* her appearance.

However, this is a world within which not only can woman's absence be apprehended and accepted, but women themselves can learn to master the gaze, and can exist entirely outside an interpretative dynamic which reads them as castrated. The problem is not being looked at and objectified, it is being *over*looked.[34] Moreoever, it is probably the *presence* of Fevvers's extra parts which needs explaining here, not apparent visual absences, and as if to consolidate the possibility of a physically and fantastically 'extra-woman' who is also the subject of the gaze, the bizarre character of Fanny Four-Eyes appears, with eyes where her nipples should be (although this is more of a curse than a gift: 'she saw too much of the world altogether' and asks 'How can you nourish a baby on salt tears?'[35]).

I do not wish to argue, however, that Carter's is a naïve feminism which has simplistically set about presenting a world within which phallic women are splendidly dominant and in visual control. Rather, this is a world within which a number of visual relationships are encountered, and specifically explored in terms of the construction of gender. *Nights at the Circus* is doing several things at once. It is keeping a number of 'truth' possibilities alive, so that the question of 'Is she fact or is she fiction?' is never resolved. The 'fantastic' moment of hesitation continues through the text, laced with the resonance of unconscious fantasy. *Nights* also asks what would happen if a woman embraced her 'freakish' visage – the image which man can only accommodate with difficulty – living the fantasy of herself in the open and enjoying the world of appearances. Early on in the novel Fevvers asserts her resistance to the charms of 'a magic prince [whose] . . . kiss would seal me up in my *appearance* for ever'. And yet, if not sealed up in her appearance, Fevvers is certainly complicit in the world of looks; in the same speech she asks, with some approval, 'is it not to the mercies of the eyes of others that we commit ourselves on our voyage through the world?'[36] Is this, then, the tale of the passive exhibitionist who can only act out her subjection to the gaze of others – the woman who needs to be seen to *be* – who needs to be seen *in order to be*? (The exhibitionist's version of Descartes, then: I *am seen* therefore I am.[37])

34 The terror of *not* being looked at is explored by Toni Morrison through the character of Denver in the 1987 novel, *Beloved* (London: Chatto & Windus, 1988). For Denver, it is not only love which resides in the other's gaze, but her own sense of identity too: 'It was lovely. Not to be stared at, not seen, but being pulled into view by the interested, uncritical eyes of the other. Having her hair examined as a part of her self, not as material or a style. . . . And to be looked at by her, however briefly, kept her grateful for the rest of the time when she was merely a looker' (pp. 118–19).

35 *Nights*, p. 69.

36 Ibid., p. 39.

37 I discuss another version of this through the female voyeurism central to D. H. Lawrence's work, in *Sex in the Head*, p. 92.

However, the novel is not primarily a shaggy-dog story exploring the nature of Fevvers's reality, but a process of visual re-education which Walser (as reader) has to undergo in relation to the impossible. The fetishist negates what he sees – he has to insert something into the gap so that he can believe it instead. Walser responds with increasing belief, maintaining his rational 'hesitation', so that what he sees is eventually exactly what is. And he falls in love with it, but this is not a love which is blind; rather, it is a love which need not disavow the impossible. In the end, Fevvers admits that she fooled Walser – that she is in control of what he sees – but still it is unclear whether she fooled him on the question of her feathers or of her virginity. He can only join in as 'the deceived husband, who found himself laughing too, even if he was not quite sure whether or not he might be the butt of the joke'.[38]

III
Primal Fantasy and the Paranoid Woman

What Walser sees and what Freud's boy sees are both fantasies, although this is not to say that they are not predicated on a real vision. Walser the journalistic fact-monger has entered the twilight zone of literary fantasy, the all-things-are-possible paradigm of magic realism. For Freud, identity begins with the unbelievable visions of primal fantasy. The truth value of neither form of fantasy is in question here: for Freud and the fantasist, truth lies in the eye of the beholder, so much so that Walser's reality may be entirely hallucinatory, the Freudian boy may never have 'actually' seen such things. Psychic and sexual life begins with a fantasy or image which may never have happened, and Walser's journey ends in a remarkably similar place.

Thus far in our discussion of *Nights at the Circus*, the novel has engaged in a kind of visual utopianism. But the Freudian subject does not inhabit the Shaman's world, in which seeing is believing in the most benign possible way. The confrontation of the Freudian child with one of Freud's primal phantasies (like the boy's confrontation of the image of a penisless woman) kicks off a process which is characterized by anxiety and negation – the process of coming into gender and subjectivity. Sexual subjectivity is also a 'fiction', based on some sort of belief in or confrontation with the visual.

Two issues which focus how psychoanalysis theorizes the subject are being addressed and reworked here, and Carter's text offers a fictional reading of the intersection of these issues. The first is the question of fantasy, and psychoanalysis's privileging of fantastic experience over (or as)

38 *Nights*, p. 295.

'the real' – its insistence that fantasy is true on its own terms. This shift from belief in the empirical 'reality' of Freud's Seduction Theory to the belief in the reality of psychic and unconscious life, the reality of desire, has already been discussed in detail, but it continues to be central. It could be said that the Shaman lives in such a psychoanalytic space, although his non-neurotic response to the policing of psychic boundaries (he seems to have none, and his anxieties are consequently minimal) also differentiates him from the space of psychoanalysis. The second issue which hovers over the passage is that of visual identity, of the self coming into being in terms of visual images. Both of these have been examined most acutely in film theory and feminism, the latter having a special interest in the politics of both pernicious and progressive uses of fantasy.[39]

Film theorists are not slow in pointing out that, for both Freud and Lacan, subjectivity begins with a vision. For Freud the 'primal fantasy' is one of the stock scenarios or formative images which govern infantile life and our memory of it. Primal fantasies are most likely to be imagined scenes not necessarily actually witnessed at all, of parental sex, of castration represented by the vision of the female genitals, of seduction; as Laplanche and Pontalis put it in *The Language of Psycho-Analysis*, primal fantasies are 'typical phantasy structures . . . which psycho-analysis reveals to be responsible for the organisation of phantasy life, regardless of the personal experience of different subjects'.[40] Laplanche and Pontalis also worked together on their seminal essay, 'Fantasy and the Origins of Sexuality' (1968), in which they discuss the knot or paradox which establishes fantasy as 'the fundamental object of psychoanalysis'.[41] For the Lacan of 'the mirror-phase' the *gestalt* image which the child identifies with in the mirror, transforming itself into a subject through that act of identification, is crucially not itself but an image of itself (the mirror may even be the mother). Seeing with a certain kind of belief through which it can recognize the image as that of itself, the child begins the process of subjectivity. Upon this range of visual images the self is built (even though none of these images need actually be 'real', only 'believed in'). Seeing the primal scene, the woman's castrated absence, and taking on board this new kind of visual knowledge, this 'belief' is at the origin of the construction of the Freudian self. Seeing is believing, even if the visual knowledge taken in by the seer is later remembered and reworked with anxiety and horror.

Freud first used the term 'primal phantasy' in the important short work of 1915, 'A Case of Paranoia Running Counter to the Theory of the Disease', written twelve years before 'Fetishism'.

39 Jeffrey Masson's anti-psychoanalytic work on Freud's abandonment of the Seduction Theory in the 1980s was welcomed by some feminists for its critique of Freud and its privileging of real trauma. See Chapter 1 for a discussion of this.
40 p. 331.
41 *International Journal of Psycho-Analysis*, 49/1 (1968), p. 7.

> Among the store of unconscious phantasies of all neurotics, and probably of all human beings, there is one which is seldom absent and which can be disclosed in analysis: this is the phantasy of watching sexual intercourse between the parents. I call such phantasies – of the observation of sexual intercourse between the parents, of seduction, of castration, and others – 'primal phantasies'.[42]

This first announcement of primal fantasy itself comes in a case history concerned with a woman as object (and subject) of the gaze. The paranoid woman in question doesn't so much believe in what she sees as what she hears: as she is making illicit love with a man in his room, she hears (or thinks she hears) the sound of a camera fixing her image in a compromising position. The hallucination (as Freud reads it) is aural, but what it betrays is an elaborate fantasy of vision. From touch (the love-making and her feelings about this – Freud comes to the conclusion that the sound of the 'camera' is actually the throb of the woman's clitoris), to sound (the click across the room), to vision (a whole story of looking at sex), a chain of sensory associations take the story back to a primal image. As Freud unravels the paranoid knot, it emerges that the woman consciously fears that she has been photographed by a man hidden behind a curtain; unconsciously she is identifying with her mother in the primal scene, with her lover playing the part of her father. Or rather, she is inserting her mother into *her* role with her lover-father, so that the position she occupies becomes that of the camera (the child witnessing the primal scene) watching the figures having sex. Whilst Freud writes that 'the part of the listener' – originally herself as a child, peeping unnoticed – 'had then to be allotted to a third person' (i.e. the imagined man with the camera who is the manifest object of her paranoia), actually it is the camera which 'sees' as she does or had done, positioned as the privileged infantile viewer of the scene. And as Freud was to elucidate in his essay of four years later, 'A Child Is Being Beaten', all good fantasies involve the splitting of the fantasizing subject into a number of identificatory positions – beaten child, beating father and voyeuristic witness of the whole scene. Indeed, in fantasy the subject may not even be present – fantasy can render the subject invisible, can un-write the process of subjectivity, situating the fantasist before, or outside of, his or her identity, and the gender alignments which subjectivity implies. According to this rationale, here the paranoid woman would fantasize the scene from (at least) three positions too: as herself in *flagrante delicto*, as her own mother in a past moment still being replayed in the present, *and* as voyeuristic camera. As Carol Clover writes of the multiple identifications of the cinema audience (here, the horror-film spectator),

42 PF vol. x, p. 154.

> [J]ust as attacker and attacked are expressions of the same self in nightmares, so they are expressions of the same viewer in horror film. We are both Red Riding Hood and the Wolf; the force of the experience, in horror, comes from "knowing" both sides of the story.[43]

However, whilst all of these positions are experienced by the paranoid woman, they are experienced painfully. We might cut across at this point to Anaïs Nin's erotic short story in *Delta of Venus*, 'The Veiled Woman', which shows a woman in complete pleasurable control, setting herself up as visual object and controlling subject. In a way the veiled woman is the opposite of Freud's paranoiac, and her role as ancestor of recent feminist celebrants of pornography and the pleasures of exhibitionism (militating against an earlier 'cultural feminism' which denied the pleasures of the gaze for women) might situate her alongside some of Carter's experimental exhibitionists.[44] Nin's woman is the opposite of Freud's, in that she consciously places herself centre-stage, orchestrating male voyeurism to facilitate unwittingly her exhibitionism. The woman of the title arranges (via a go-between) for a man to pay out $100 secretly to watch whilst another man has sex with her (unaware that they have an audience), for (what turns out to be) $50 of the money the first man has given her. The money, then, is passed from one man to another, with the woman in the middle getting the pleasure of sex, the pleasure of being seen, and a neat profit of $50.[45] The click of a camera would only heighten her *jouissance*; in a sense, this is a woman entirely absorbed by the Lacanian gaze as Other (discussed in Chapter 2), as she received the 'gleam' of the glance with pleasure.

Not so Freud's paranoiac, but the comparison with Nin's wanton woman demonstrates how these different forms of fantasy are beginning to fictionalize – reread at a later stage, with anxiety or desire – a more primal scenario. I have said that Carter's story is hooked on a moment of 'hesitation' within two knowledge systems (the generic and the subjective, asking both 'what is going on?' through the literary fantasy and 'who is she?' through a fantasy of identity). In a similar way, Freud's paranoiac straddles the line between psychic (primal) fantasy and sexual (pornographic) fantasy – this is a scene which plays out an unconscious scenario which still demands to be reworked, within the terms of an erotic *vignette*. The paranoid woman resists, but read another (Nin's)

43 Carol J. Clover, *Men, Women and Chain Saws* (London: BFI, 1992), p. 12.

44 Indeed, 'The Veiled Woman' was the piece of 'classy erotica' which was reprinted alongside the male nudes in the first issue of *For Women*, the first of a spate of women's porn which appeared in 1992 in Britain, demonstrating that the porn industry had woken up to the fact that there is a market for such magazines, as well as underlining the relationship between this text and the anticipated visual pleasures of the surrounding images.

45 Jacqueline Rose reads the story as a parable of the whole apparatus of cinema in her essay 'Woman as Symptom' in *Sexuality in the Field of Vision*.

way, the scene becomes quite different. The line the paranoiac treads between fantasy as pleasure or horror is a thin one: viewed from the position of the veiled woman, Freud's patient's situation is highly desirable. In a sense the two could be mapped onto and read as versions of the Sadeian couple, Justine and Juliette, with Nin's woman (like Juliette) experiencing with delight everything that Freud's woman (like Justine) fears and loathes. The paranoiac's vision thus works on at least two levels: it is charged with, and originates in, the power of the primal fantasy or 'vision' of sex, and it has the narrative structure of a classic sexual fantasy (which Freud seems to have missed altogether), a scene viewed from a number of subject positions.

As Rosemary Jackson reminds us, the word 'fantasy' is derived from the Latin *phantasticus*, which means visible, visionary *and* unreal – a very wide definition which takes in all of these sights or sites of analysis. For each of the cases I have touched upon, the construction and the breakdown of identity takes place in the realm of visual fantasy. Freud's model indicates that explicitly for males, and implicitly for females (who also pass through the castration complex as well as experiencing the pervasive power of primal fantasy), sexual identity is built on a visionary experience of impossibility. I want to argue that the fantasy-moment of 'hesitation', which may be one way of reading the fetishist's trauma, or the fantasy-reader's uncertainty (which comes before an explanation is established for the impossible thing seen), constitutes a challenge to the construction. Carter keeps the 'hesitation' going for the whole novel for precisely this reason, extending the gap between possible explanations or definitive answers to the question 'Is she fact or is she fiction?' The fetishist responds in two different ways: he tries to close the abyss which is opened up by the woman's castrated 'absence' as fully as possible (inserting an object into it), but as long as the fetish remains, it must stand as much for the awful absence as for its disavowal.

IV
The Creative Lives of Objects: Winnicott and Klein

I now want to shift across to quite a different psychoanalytic tradition, which also looks back to the fantasies of very early life in order to return to Carter and examine her through a different psychoanalytic lens. Carter's story makes Fevvers the impossible object, whose body opens up a protracted experience of 'hesitation' on the part of readers or viewers, inside the text and out. Although my discussion so far has highlighted the Freudian and Lacanian processes of analysis focusing on the self's strata, its repressions and structures developed through a complex internal history of individual sexuality, another post-Freudian strand looks at how the child develops in terms of the world, its relationship with objects, its individuation from its mother. What has come to be known as 'object-relations theory' (influenced originally by the work of Melanie Klein, and developed through the work of John Bowlby, Harry Guntrip and D. W. Winnicott) has also engendered a significant history of cultural, and specifically feminist, criticism. I want briefly to look at how these analysts might offer an alternative view of fantasy, creativity, and the construction of the subject through social relationships rather than internal drives, before returning to the operation of these in Carter's work.

Both Klein and Winnicott worked with very young children (unlike Freud, who looked for the child *in* the adult), and Klein developed a method of analysing preverbal children through play, which focused on the infant's experience within the mother–child dyad, and particularly its formative relations with objects which come to be perceived as separate from the self. Klein rewrites the Oedipal drama to make the mother and her body (as prime object for the child) the central figure in infant development. Her 'play technique' encouraged the child to enact scenarios, narratives, fantasies, with objects (neutral toys provided for the analysis) which were then interpreted in a similar way to the interpretation of linguistic phenomena and dreams in adult and classic Freudian analysis. The toys allow the child to play out its fantasies actively, developing a symbolic system through the objects themselves: 'Klein moves the symbol into the consulting room and offers it to the child, senses its anxiety and discovers its fantasies,' writes Juliet Mitchell.[46]

Through this process Klein constructed an alternative view of early psychic development, based on the strong motivating positive and negative emotions which the child projects onto objects, starting with the breast. These responses to objects, particularly the breast, are then understood through a post-Freudian model of the drives (or instincts). By the 1920s (when Klein started her work) Freud's theory of the instincts (or drives) was predicated on a balance or duel between life and death instincts. In 'Analysis Terminable and Interminable' (1937) he writes,

> Only by the concurrent or mutually opposing action of the two primal instincts – Eros and the death-instinct - never by one or the other alone, can we explain the rich multiplicity of the phenomena of life.[47]

The Kleinian child negotiates its relationship with the world through a similar struggle of opposing instincts – life and death instincts. For Klein, the infant does not simply repress aspects of its growing desire and traumatic relationship with the world. In her lucid discussion of the basic

46 Introduction to *The Selected Melanie Klein* (Harmondsworth: Penguin, 1986), p. 23.
47 *SE* vol. xxiii, p. 243.

structures of Kleinianism, Mitchell singles out four central mechanisms in which the child might engage to deal with its unwelcome perception of reality: splitting (of the ego as an act of self-protection, or by the ego of an object perceived as both good and bad), projection (of the ego's feelings onto the object), introjection (taking into the ego feelings which the ego associates with the object), and projective identification (projecting feelings onto the object, which the ego then identifies with and fantasizes about).[48] The infant ego's understanding of reality is then constantly worked on in early psychic life, and these responses are consolidated into two dominant positions (as distinct from the 'complexes' or 'phases' of Freud and Lacan – Klein's 'positions' are maintained throughout adult life as part of the individual's way of responding to the world). The Kleinian child enters the world already charged with the passions of love and hate. In 'Notes on Some Schizoid Mechanisms' (1946) Klein writes:

> object relations exist from the beginning of life, the first object being the mother's breast which to the child becomes split into a good (gratifying) and bad (frustrating) breast; this splitting results in a severance of love and hate.[49]

Klein thus strengthens the Freudian duality of life-instincts and death-instincts, or Eros and Thanatos, which become the dominant principles from birth. Objects which the child encounters are experienced through these polar emotions. The breast in particular is the first object, split by the baby through its ambivalent feelings into a good and bad object, but this in itself produces intense anxiety. The baby

> fears that the object on which it vents its rage . . . will retaliate. In self-protection it splits itself and the object into a good part and a bad part and projects all its badness into the outside world so that the hated breast becomes the hateful and hating breast.[50]

This is the paranoid-schizoid position, and is complemented by the depressive position, which is characterized by guilt felt in relation to these prior negative and anxious emotions.[51] The 'fusion of erotic and destructive impulses' are, for both Klein and Winnicott, 'a sign of health'.[52] In each case, psychic development takes place in relation to the

48 See Mitchell's introduction to *The Selected Melanie Klein* for a fuller discussion of these concepts.
49 *The Selected Melanie Klein*, p. 176–7.
50 Mitchell, Introduction, p. 20.
51 For a succinct discussion of the paranoid-schizoid position, see Klein's 'Notes on Some Schizoid Mechanisms' (1946), pp. 175–200 of *The Selected Melanie Klein*; for a discussion of Klein's conception of the depressive position, see 'A Contribution to the Psychogenesis of Manic-Depressive States' (1935), also in *The Selected Melanie Klein*, pp. 115–45.
52 *Playing and Reality* (1971) (Harmondsworth: Penguin, 1988), p. 82.

baby's growing awareness of objects which are increasingly separated from its own body.

D. W. Winnicott builds upon this a theory of individuation and creativity also bound up with the processes of play and object-related fantasy. For him, there is a direct relationship between infant fantasy and play, and adult creative and cultural life, mediated by the 'transitional object'. What the object is (a teddy bear, a blanket, a thumb) is irrelevant – what is important is its function in drawing the child out into a relationship with the world. Here Winnicott takes object-relations theory away from Klein's emphasis on fantasy and primal drives, into a developing discussion of the child's *real* relationship with the social world mediated by *real* objects. The child is motivated not just by drives but by relationships, and objects mediate these relationships. In object-relations theory, fantasy is staged through conscious interrelationships, and not only in an internal, private theatre. What is important for our discussion here is that Winnicott focuses on how the child plays with the object as an embryonic form of cultural relationship. The transitional object is the child's 'first non-me possession', through which it learns symbolism and first experiences play. The object mediates the child's first relationship with the external world, marking out a space for play between self and other (and at the same time marking the difference *between* self and other), a space which culture is to occupy.

In many ways the break which object-relations theory makes with classic Freudian models means that Winnicott's position on the entry into culture is incompatible with the Freudian and Lacanian readings I am offering elsewhere in this book. However, in its emphasis on and return to the very early phase of psychic development which Freud identified as the pre-Oedipal moment of unity with the mother and the figure of the mother herself (before the intrusion of the father), object-relations theory has influenced feminist psychoanalytic readings since the 1970s, particularly of women writers. In addition, Winnicott does offer a model of the subject's relationship to culture which we should pause on. In 'The Location of Cultural Experience', one of the essays included in the important collection *Playing and Reality* (1971), Winnicott accounts for a creative space in the self which, he argues, does not figure in Freud:

> Freud did not have a place in his topography of the mind for the experience of things cultural. He gave new value to inner psychic reality, and from this came a new value for things that are actual and truly external. Freud used the word 'sublimation' to point the way to a place where cultural experience is meaningful, but perhaps he did not get so far as to tell us where in the mind cultural experience is.[53]

53 Winnicott quoting himself in 'The Location of Cultural Experience', *Playing and Reality*, p. 112.

Winnicott is then proposing a topography which accounts for how cultural experience is opened up within the subject, whereas Freud's theory of sublimation discusses artistic drives (as well as other conscious phenomena such as character traits[54]) as a distorted or refined cultural expression of sexual drives. According to Freud, the pressures of civilization cause the energy of infantile sexual impulses, if 'diverted, wholly or in great part, from their sexual use and directed to other ends',[55] to be reformed and expressed differently in later life in culturally appropriate ways, such as artistic creativity. In 'Civilisation and its Discontents' (1930 [1929]) he writes that 'work is desire held in check', and in his theory of sublimation, developed across the range of his work, he suggests that cultural practices in particular are desire held in the 'check' of sublimation. In 'The Tendency to Debasement in Love' (1912), for instance, he writes:

> The very incapacity of the sexual instinct to yield complete satisfaction as soon as it submits to the first demands of civilization becomes the source, however, of the noblest cultural achievements which are brought into being by ever more extensive sublimations of its instinctual components.[56]

Winnicott, as we shall see, accounts for cultural acquisition, creativity and appreciation spatially, in terms of an early conscious relationship with objects, whereas Freud explains it instinctively, as a diverted form of an unconscious, repressed Eros. In 'The Resistances to Psychoanalysis' (1925 [1924]), discussing the severe opposition psychoanalysis received in its suggestion that 'art, religion and social order originated in part in a contribution from the sexual instincts', Freud writes:

> Psychoanalytic theory maintained that the symptoms of neuroses are distorted substitutive satisfactions of sexual instinctual forces, the direct satisfaction of which has been frustrated by internal resistances. Later on, when analysis had extended beyond its original field of work and began to be applied to normal mental life, it sought to show that these same sexual components, which could be diverted from their immediate aims and directed to other things, made the most important contributions to the cultural achievements of the individual and of society.[57]

54 See in particular 'Character and Anal Eroticism' for a discussion of how 'character in its final shape is formed out of the constituent instincts: the permanent character-traits are either unchanged prolongations of the original instincts, or sublimations of those instincts, or reaction-formations against them' (PF vol. vii, p. 215).

55 Freud, *Three Essays on the Theory of Sexuality*, ibid., p. 94.

56 Ibid., p. 259.

57 PF vol. xv, pp. 269 and 268.

The problems Freud encountered in suggesting this largely rested on the *decentring* effect of this model, which proposed the translation of one (lower) drive into another (higher) cultural form; even our most refined, cerebral, civilized achievements are predicated on, indeed are *forms of*, sexuality 'sublimated' into a socially acceptable shape. As Freud put it, 'The ruler's throne rests on fettered slaves': 'Human civilization rests upon two pillars, of which one is the control of natural forces and the other the restriction of our instincts.'[58]

For Winnicott, however, cultural experience emerges from a process which is primarily interactive, both because the subject is developed interactively (through the interconnection of people, predicated upon an experience between self and object) and because creativity takes place within a history (through interaction with a set of cultural 'givens' – a tradition). Thus one experiences culture through a dynamic of originality and tradition (both terms under severe scrutiny in literary studies):

> *[I]t is not possible to be original except on a basis of tradition.* . . . The interplay between originality and the acceptance of transition as the basis for inventiveness seems to me to be just one more example, and a very exciting one, of the the interplay between separateness and union.[59]

Adult cultural experience is then a form of the interaction between self and other, union (with the mother/object) and separateness (individuation), formed through early play. The transitional object is a prototype cultural object, or rather, the act of playing with it opens up the space which cultural relations will come to occupy. 'Culture' is, then, both an effect of the cultural objects, attitudes and forms existent in the world and the space within the subject which is open to 'receive' it:

> I have used the term cultural experience as an extension of the idea of transitional phenomena and of play without being certain that I can define the word 'culture'. The accent indeed is on experience. In using the word culture I am thinking of the inherited tradition. I am thinking of something that is in the common pool of humanity, into which individuals and groups of people may contribute, and from which we may all draw *if we have somewhere to put what we find.*[60]

The problem with this, however, is the same problem inherent in the model of self and other which Winnicott uses to describe individuation – that self and other, inside and outside, exist as preformed identities. The question of whether they come into being through their mutual interaction is answered with reference to the paradox of the origin of the object itself:

58 Ibid., p. 269.
59 *Playing and Reality*, p. 117; his italics.
60 Ibid., p. 116; his italics.

> the essential feature in the concept of transitional objects and phenomena . . . is *the paradox, and the acceptance of the paradox*: the baby creates the object, but the object was there waiting to be created and to become a cathected object.[61]

For Freud (and Klein), the anxious negotiation of the world and fantasies regarding the world construct the sexual subject, and boundaries between inside and outside are formed through the process of working on the world. Winnicott's subject faces the problem of how to separate itself from its mother, yet it also enters the world with an integral sense of internal identity, a prototype ego: as he writes in conclusion to 'Creativity and its origins', 'After being – doing and being done to. But first, being.'[62] Inside and outside are preordained realms which the child negotiates more or less successfully via the transitional object which 'is *not an internal object* (which is a mental concept) – it is a possession. Yet it is not (for the infant) an external object either.'[63]

What becomes of the object is that it is replaced by culture:

> Its fate is to be gradually decathected, so that in the course of years it becomes not so much forgotten as relegated to limbo. By this I mean that in health the transitional object does not 'go inside' nor does the feeling about it necessarily undergo repression. It is not forgotten and it is not mourned. It loses meaning, and this is because the transitional phenomena have become diffused, have become spread out over the whole intermediate territory between 'inner psychic reality' and 'the external world as perceived by two persons in common', that is to say, over the whole cultural field.[64]

Winnicott continues, 'at this point my subject widens out'; not only does it encompass creativity, dreaming, religion, but also fetishism.[65] In 'The Location of Cultural Experience' he famously writes:

> The place where cultural experience is located is in the potential space between the individual and the environment (originally the object). The same

61 Winnicott, 'The Use of an Object and Relating Through Identifications', in *Playing and Reality*, p. 104; his italics.

62 p. 99.

63 Winnicott, 'Transitional Objects and Transitional Phenomena', p. 11; his italics. See also Elizabeth Wright's discussion of this and of Alfred Lorenser and Peter Orban's critique of Winnicott, in *Psychoanalytic Criticism*, pp. 94–7.

64 Winnicott, 'Transitional Objects and Transitional Phenomena', p. 6.

65 Here fetishism is not explicitly linked to castration anxiety, as is Freud's discussion of the fetish object. Winnicott writes: 'The transitional object may eventually develop into a fetish object and so persist as a characteristic of the adult sexual life' (ibid., p. 10). Fetishistic experience is thus also read as a form of 'culture' – a play-relation to objects standing in for, and facilitating, a relationship with something else. The separation or displacement of one object from another which the fetish comes to symbolize is at the heart of both processes under discussion here, even if the psycho-sexual mechanisms engendering each come from distinct psychoanalytic traditions.

> can be said of playing. Cultural experience begins with creative living first manifested in play.[66]

What is important here is that the play-space Winnicott marks out is one in which, as long as it is kept open, any number of imagined, contradictory experiences can prevail. In Meredith Ann Skura's *The Literary Use of the Psychoanalytic Process* Winnicott's model is interrogated to figure the fantasy-past of author as well as reader. In an intriguing analysis of Dickens's *The Old Curiosity Shop* read as fantasy, she argues that fiction 'is a "transitional" world that is neither a part of ordinary reality nor a mere falsehood but which has its own provisional truth'.[67] At only one point in Dickens's novel, however, do

> fantasies meet the real world on a ground midway between them, and creations are known and accepted for what they are: neither real nor false but something different and better.[68]

This moment of play (or – to turn to our earlier term – of fantastic hesitation) focuses in *The Old Curiosity Shop* on the character of Swiveller, who 'is the only one of the characters who lives in a fiction known and relished as a fiction'.[69] A reading of Carter's *Nights at the Circus* could be constructed, however, which would situate the whole text at this middle point of play, at which fiction is precisely 'relished' *as* a fiction.

Until fairly recently, any remarriage of Freud and object-relations theory was untenable for purists, who have emphasized the radical differences which have opened up between the two traditions since Klein took her work in new directions in the 1930s. A recent feminist 'return to Klein'[70] has followed the increased interest in non-Kleinian object-relations theory and in the power and potential of the pre-Oedipal which took place in the 1970s and 1980s, particularly in America. One writer who has sought to draw these traditions together is American feminist writer and analyst Jessica Benjamin, and it is through her work that I wish to return to Carter. Benjamin reads Winnicott through Hegel and Habermas (particularly to justify the assumption that 'life begins with an emergent awareness of self and other'[71]), in order to account for the problems and pleasures of violent or unequal sexual power relations. Her

66 *Playing and Reality*, p. 118.
67 *The Literary Use of the Psychoanalytic Process* (New Haven, CT, and London: Yale University Press, 1981), p. 197.
68 Ibid., p. 198.
69 Ibid.
70 In particular, the recent work of Juliet Mitchell and the discussion of Klein in Jacqueline Rose's *Why War? Psychoanalysis, Politics and the Return to Melanie Klein*, (Oxford: Blackwell, 1993). This 'return to Klein' takes a different route from that taken by the Chodorow-inspired feminism of the 1970s and 1980s explored below.
71 Jessica Benjamin, *The Bonds of Love* (London: Virago, 1988), p. 36.

question addresses not primarily how successful individuation from the mother takes place, but how the child comes to form 'healthy' (as opposed to sado-masochistic or dominating/dominated) relationships with others:

> Once we accept the idea that infants do not begin life as part of an undifferentiated unity, the issue is not only how we separate from oneness, but also how we connect to and recognize others; the issue is not how we become free of the other, but how we actively engage and make ourselves known in relationship to the other.[72]

If classic Freudianism is concerned with processes taking place between different strata of the self (intrapsychically), Benjamin forms a psychoanalytic social theory by supplementing this with a stronger sense of how that self interacts socially. Relationships, and the self itself, are forged through the 'intersubjectivity' of the child in the social. Rather than synthesizing the two traditions, one approach complements the other.[73] A feminist account of problems of domination thus requires an account of the intersubjective construction of the subject:

> Whereas the intrapsychic perspective conceives of the person as a discrete unit with a complex internal structure, intersubjective theory describes capacities that emerge in the interaction between self and others. . . . The crucial area we uncover with intrapsychic theory is the unconscious; the crucial element we explore with intersubjective theory is the representation of self and other as distinct but interrelated beings.
>
> I suggest that intrapsychic and intersubjective theory should not be seen in opposition to each other (as they usually are) but as complementary ways of understanding the psyche.[74]

But as I have indicated, Benjamin's concerns are dark ones, to do with the psychic origins of power, authority, and sexual violence. If Klein opens up the tracts of destructiveness in the infant's fantasy-life – its aggressivity, its 'epistemophilia', its death drives – Winnicott pinpoints the moments of specific fantasy-destruction in the child's negotiation of the object to which it has been related, and here Benjamin's interest is especially keen. I discussed Carter's fantasy as a moment of hesitation which concerns the sexual reality of the subject, and the reader's ability to assimilate this 'reality' even if it might appear impossible – so the reader (Walser or us) is 'infected' by the fantasy-image she or he witnesses, and must hesitate between possibilities (or resolve the dilemma

72 Ibid., p. 18.
73 Ibid., p. 251 n.
74 Ibid., p. 20.

as the fetishist does in a specific context). Benjamin reads Winnicott's 'The Use of an Object and Relating Through Identifications' for the light it sheds upon the role of fantasy-*de*struction in the *con*struction of the subject. Winnicott's child destroys the object in fantasy, in order to see it survive in reality, moving through three positions:

> [A]fter 'subject relates to object' comes 'subject destroys object' (as it becomes external); and then may come '*object survives* destruction by the subject'. But there may or may not be survival. A new feature thus arrives in the theory of object-relating. The subject says to the object: 'I destroyed you', and the object is there to receive the communication. From now on the subject says: 'Hullo object!' 'I destroyed you.' 'I love you.' 'You have value for me because of your survival of my destruction of you.' 'While I am loving you I am all the time destroying you in (unconscious) *fantasy*.' Here fantasy begins for the individual. The subject can now use the object that has survived.[75]

Clearly, the distinction has to be made here between fantasy for Winnicott which begins at this late stage in development, and fantasy in Freudian psychoanalysis, which the child is born with and into (primal fantasy). What is important for Benjamin is the way in which the object becomes a focus for the child's move into a relationship with the world in a number of ways – it is only through successful fantasy-interaction and symbolism ('pretend' scenarios) that 'healthy' relationships can develop. Benjamin moves on to a close literary analysis of an 'unhealthy' scenario, Pauline Réage's 'classic' piece of sado-masochistic erotica, *The Story of O*. This is a fantasy in more ways that one, within which 'I destroyed you' does not necessarily resolve into 'I love you' or 'You have value for me because of your survival', but it has clear parallels with both Freud's 'Case of Paranoia' and Nin's 'Veiled Woman' in its conflation of masochism with exhibitionism. There is also a connection to be drawn between Benjamin's analysis of domination and Angela Carter's rather different 1979 analysis of the same area in *The Sadeian Woman: An Exercise in Cultural History*. Although there is not the space to open this issue up here, I will return to Carter's discussion of Sade's peculiar rendition of a Female Oedipus in a moment. Of more general interest is Benjamin's desire to draw Freud, Winnicott and feminist literary analysis together, and I now want briefly to outline other ways in which object-relations influenced criticism developed in the 1980s.

75 Winnicott, 'The Use of an Object. . .', pp. 105–6; his italics.

V

Mothers and Daughters: Pre-Oedipal and Other Feminisms

In Chapter 1 I noted Freud's 'blindnesses' in the 'Dora' case, and one of these (which has formed an important focus for feminist critique) is his marginalization of the homosexual desire between Dora and Frau K. If Frau K. is represented for Dora as a maternal figure, this means that (for Freud) her role is to inspire jealousy in Dora, who sees her as a rival in her love for her father. Dora's love for, and identification with, the woman-as-mother is here read by Freud only in terms of Oedipal jealousy. Later in his work, however, he was to find a quite different way of understanding female desire, and particularly women's connections with the mother and her psychic representatives. I now want to turn to this alternative patterning of female desire and identification. By also shifting back to an earlier moment as its primary focus – the pre-Oedipal bond between mother and infant – object-relations theory, as well as certain strands of contemporary French feminism, have focused on the role of the mother rather than that of the father and the law of patriarchy in the development of femininity. Having maintained for much of his work a basic model of female development which posited femininity as the mirror-image, the reverse, of the male Oedipal model, Freud makes a famous about-turn in 1931, when he recognizes that femininity may not be as simply accountable to Oedipus as he had previously thought. In the essay 'Female Sexuality', and then a little later in the *New Introductory Lectures* (1933 [1932]), Freud acknowledges that simple symmetry between the sexes cannot exist ('[w]e have . . . long given up any expectation of a neat parallelism between male and female sexual development'[76]). The little boy's Oedipal attachment to the mother is understandable enough, since

> '[h]is first love-object was his mother. She remains so; and, with the strengthening of his erotic desires and his deeper insight into the relations between his father and mother, the former is bound to become his rival.[77]

But the little girl also starts with the mother as her primary love-object, and this original bisexuality makes the translation of the Oedipal model less clear, especially since many women remain 'arrested in their original attachment to their mother'. 'This being so', he writes, 'the pre-Oedipus phase in women gains an importance which we have not attributed to it hitherto', possessing 'a far greater importance in women than it can have

[76] 'Female Sexuality', PF vol. vii, p. 372.
[77] Ibid., p. 371.

in men'.[78] He then goes on to make a famous analogy, which has acted both as a challenge to feminists in recent years and as something of an admission that he is vacating the field. Freud often discusses early life as a kind of subjective prehistory (or rather, a prehistory of the self which is also pre-subjective), and here he offers another narrative-within-a-narrative, the buried myth at the root of femininity:

> Our insight into this early, pre-Oedipus, phase in girls comes to us as a surprise, like the discovery, in another field, of the Minoan-Mycenaean civilization behind the civilization of Greece.[79]

The revision is crucial: acknowledging that the girl does not emerge heterosexually oriented towards the father, but that she begins life in a relationship of unity with the mother as original love-object, Freud admits that her primary connection is with a member of the same sex. How Freud proceeds to solve what he calls a little later, in the *New Introductory Lectures*, 'the riddle of femininity'[80] is not my concern at present; rather, I am interested in the opening this question has offered to other writers more recently. The return to the mother is, then, starting-point for (at least) three different traditions of feminist thought. In America, the work of Nancy Chodorow, who marries a social construction of the female subject with a form of non-Kleinian object-relations theory, has had a profound influence on humanist feminist criticism, to which I will turn in a moment. At the same time, work in French feminism, on the archaic mother and the force of the pre-Oedipal, has developed into two different bodies of thought, converging around distinct critiques of Oedipus which takes place alongside a rediscovery of different routes back to the mother – both Hélène Cixous and Luce Irigaray have developed very different emphases, on the mother as creative source and inspirational figure of plenitude, and as estranged figure cut off from the daughter by patriarchy. In addition, Julia Kristeva has reworked Freud's pre-Oedipal moment of unity with the mother in other, highly significant ways.

Firstly, both Irigaray and Cixous valorize the daughter's relationship with her mother but focus on different aspects of this connection. For Irigaray, patriarchy (through its conception of women as castrated) serves to separate women from an essential interconnection which is potentially threatening, and, in Clare Buck's words, 'women have been systematically denied the use of the mother to fantasise their origin as

78 Ibid., pp. 372 and 377.
79 Ibid., p. 372.
80 PF vol. ii, p. 149.

feminine subjects'.[81] Despite charges of essentialism, however, Irigaray is not always engaged in a blind celebration of essential mother–daughter unity. (The lyrical poetic essay 'And One Doesn't Stir Without the Other', for instance, depicts a difficult version of the notion of fluid boundaries between mother and daughter, against which the daughter must struggle if she is to gain any form of independence.) Cixous also identifies a return to the mother as crucial to the process of reclaiming women's creative powers (which is itself connected to her work on the revolutionary role of hysteria, discussed in Chapter 1). In her famous essay of 1975, 'The Laugh of the Medusa', Cixous evokes a creative power into which women can tap if only they can relocate that 'first attachment' to the mother within the self, writing 'in that good mother's milk. She writes in white ink'. Writing the (maternal) body, through a fluid organic process which literalizes 'expression', thus heals the gap opened up between mother and daughter by the intrusion of patriachy: 'Even if phallic mystification has generally contaminated good relationships', she writes, 'a woman is never far from her "mother".'[82]

Julia Kristeva's emphasis is different. As a semiotician, critic and practising psychoanalyst, she retains elements of Freudian and Lacanian thought in her radical revision of the development of the sexual subject. Instead of engaging in a critique of the boundaries which have split women off from the pre-Oedipal in order to break them down (thus ostensibly allowing women *back* into a utopian space of essential creativity), she theorizes the pre-Oedipal ('anterior to the "mirror stage"'[83]) as an alternative poetic-linguistic source, reworked as what she calls the 'semiotic'. She uses this term to identify the phase of mother–infant fusion characterized by *jouissance* and polymorphous perversity, before the child enters the Symbolic, and before it can conceive of law or taboo (Kristeva also refers to this space as the *chora*). The rhythmic, heterogeneous impulses of the infant's bodily rhythms, which have free play in this moment prior to the intrusion of the Symbolic, suggest a way of understanding what is at stake, not just in poetic language and avant-garde communication, but in the ways in which speech breaks down at

81 Claire Buck's interrogation of Irigaray's work on the mother through an intricate reading of the poet H.D. offers a good example of the ways in which Irigaray's work has found a focus in feminist literary analysis. See Buck, '"O Careless, Unspeakable Mother": Irigaray, H.D. and Maternal Origin', in *Feminist Criticism: Theory and Practice*, ed. Susan Sellers (Hemel Hempstead: Harvester-Wheatsheaf, 1991), p. 142.

82 'The Laugh of the Medusa', in *New French Feminisms* ed. Elaine Marks and Isabelle de Courtivron (Brighton: Harvester, 1981), p. 251. Both Cixous and Irigaray have been charged with various forms of sexual essentialism; see in particular Ann Rosalind Jones, 'Inscribing Femininity: French Theories of the Feminine', in *Making a Difference: Feminist Literary Criticism*, ed. Gayle Greene and Coppélia Kahn (London: Methuen, 1985), pp. 80–112.

83 *Desire in Language* (Oxford: Basil Blackwell, 1984), p. 134.

moments of adult crisis, as well as a more generalized unconscious strain in discourse itself. '[T]here is within poetic language (and therefore, although in a less pronounced manner, within any language)', she writes in *Desire in Language*, 'a *heterogeneousness* to meaning and signification', a heterogeneousness which, she continues, is 'detected genetically in the first echolalias of infants as rhythms and intonations anterior to the first phonemes, morphemes, lexemes, and sentences'.[84] This 'uncertain and indeterminate articulation . . . does not yet refer (for young children) or no longer refers (in psychotic discourse) to a signified object'. Kristeva goes on to link it to the maternal:

> Plato's *Timeus* speaks of a *chora* . . . receptacle . . . unnamable, improbable, hybrid, anterior to naming, to the One, to the father, and consequently, maternally connoted to such an extent that it merits "not event the rank of syllable."[85]

The connection of child with mother thus facilitates, for Kristeva, an alternative, creative form of pre-language which bypasses the strictures of the Symbolic. Although she does not privilege the girl-child's access to the semiotic, others have taken this emphasis on an early, playful and heterogeneous language, as a further valorization of the creative (and subversive) role of the mother in the cultural development of the child: because of its formation *before* the entry of the Father and his Law, this alternative language continues to leave its trace and to unconsciously disrupt adult discourse.

But perhaps more influential in American feminism has been Nancy Chodorow's work, crystallized in her important 1978 text, *The Reproduction of Mothering: Psychoanalysis and the Sociology of Gender*. Here Chodorow emphasizes the mother–child dyad as the crucial formative moment in the development of the female subject (the fact that the female child is nurtured by a member of her own gender), where Freudian and Lacanian models emphasize the Oedipal and castration complex axis (which focus on the intrusion of the father), as well as the mirror phase. For Chodorow, the effect of female-dominated child-rearing practices on how the individual is structured explains how ways of mothering are 'reproduced' in daughters, as well as how girls accept their gender through their primary identification with their mothers. Here, then, we move away from fantasy towards sociology, and shift towards 'real' relations between real mothers and infants. The mother of unconscious fantasy has, for the infant, a mobile and fluid significance; the mother in Chodorow is finally a fixed, sociologically verifiable, real individual with whom the infant has concrete, conscious relations. If Freudians have argued that the former model of mobile significance has more flexibility

84 Ibid., p. 133.
85 Ibid.

in cultural analysis, humanist feminism has found the social emphasis of the latter formulation persuasive. Because of this overt stress on the construction of the individual through interactive social relationships rather than in terms of the internal dynamics of fantasy and drives, Chodorowian object-relations theory has proved seductive for some feminists. It is also a very real route back to a moment which privileges the mother; Judith Kegan Gardiner writes in her survey of 'psychoanalytic mother–daughter theory':

> One reason some feminists critics prefer psychoanalytic gender theories based on mother–daughter bonding to those based on phallic lack is that such theories permit them to disregard difference altogether. Theories of interdependence among women allow theoretical independence from men and break with past theories that define the female by deviations from the male.[86]

Yet the way in which this has been taken up in Chodorow-inspired work has often been through an emphasis on the fluid feminine identities and unclear ego-boundaries which are produced in the context of women-centred, same-sex nurturing. Due to the specific quality of this, Chodorow argues, mothers foster expressive skills in their daughters, and a distinctive, fluid form of feminine subjectivity. This has fed into a theory of an innately communicative, 'osmotic' connection between women, and a way of reading (and privileging) representations of mothers, daughters and the relationships between them as a cultural heritage which has been hitherto ignored in mainstream (for which read *male*) criticism. Also important in this development was Adrienne Rich's 1976 text, *Of Woman Born*, but a wealth of critical texts focusing on the literary mother appeared in the late 1970s and 1980s, making this almost a branch of feminist analysis in its own right.[87] Clearly, the very different routes through which the return to the mother has taken place all have distinct starting-points – Chodorow's object-relations inspired analysis has a quite different post-Freudian emphasis from the work on the pre-Oedipal and maternal creativity in Kristeva and Cixous – but feminist criticism is adept at appropriating its models from very different theoretical sources. Celebrants of this matri-centred (or even matriarchal) feminism include a line (a matrilineage) following from Virginia Woolf's famous dictum, that 'we think back through our mothers if we are women', taking in Sandra Gilbert and Susan Gubar's alternative conception of women's literary history in *The Mad Woman in the Attic*, as well as Alice Walker's image of a specifically black female creative tradition, passed down from mother to daughter around the edges of what the

86 'Mind Mother: Psychoanalysis and Feminism', in *Making a Difference*, ed. Greene and Kahn, p. 136.
87 For good surveys of object-relations-inspired feminist criticism, see ibid., and Marianne Hirsch's 'Mothers and Daughters: Review Essay', *Signs*, 7/1, pp. 200–22.

rules of slavery allowed.[88] Chodorow's account of how gender is constructed socially has, then, provided a neat critical model for how meanings are constructed textually, and how women's (literary) histories are written. For feminist critics taking up this cue, motherhood has become an important motif on the level of plot and characterization, as a way of thinking through alternative processes of creativity, and as a way of thinking through an essentially female mode of thought.[89]

VI
Sexual Fictions: Carter Makes Up

We seem to be a long way from Carter and the literary fantastic, in these meditations upon identities which have such a biographical and sociological resonance for many feminist writers. I have moved from the infant's fantasies of the mother in its pre-Oedipal unity with her to a reclamation of the mother as an alternative source of feminist psychoanalytic representation and inspiration. Parveen Adams puts the question of how feminism muses 'on the relation of mother and daughter' differently; for her, 'feminism concerns the distinction between Woman and Mother'.[90] This might also be Carter's interest too. A simple way back to her work can be found through a short examination of how she also theorizes the daughter's myth of her origins, the way in which the subject 'makes itself up'.

In a sense, Carter's writing can be seen as a move from initially figuring how women emerge into, and resist, the Oedipal family to, in her later works, an alternative, playful rendering of a more fantastic feminine construction *elsewhere*. Indeed, throughout her writing she continues a developing discourse on mothering, as even a cursory look at her corpus demonstrates. First comes a complex encounter with the erotic power of classic Oedipus, experienced from the girl's point of view (which is never that of the victim) in the early novels: *The Magic Toyshop* (1967) (with the relationship played out between Melanie and Uncle Philip), *Heroes and Villains* (1969) (which has Marianne passed from her Professor father to her savage lover Jewel), and in *The Infernal Desire Machines of Doctor Hoffman* (1972) (focused around the axis of Albertina and her father, the

88 I have discussed this emphasis on a textual matrilineage, which seeks to construct an alternative feminist tradition at the same time as it looks to an essential non-mediated connection between mothers and daughters in 'Happy Families? Feminist Reproduction and Matrilineal Thought', in *New Feminist Discourses*, ed. Isobel Armstrong (London: Routledge, 1992).

89 See Sara Ruddick, 'Maternal Thinking', *Feminist Studies*, 6/2 (1980), pp. 342–67.

90 'What Is a Woman? Some Psychoanalytic Dimensions', in *Women: A Cultural Review*, 1/1 (April 1990), p. 38.

Doctor). As her work in the 1970s proceeds, however, Carter's emphasis changes, from classic Oedipus to 'negative Oedipus' (the term used to characterize the infant's incestuous desire for the parent of the *same* sex), from connection with the father to identification with the mother. But here Carter is not so much interested in representing this as in analysing the variety of ways in which mothers and daughters can fantasize each other in print.

It would be easy enough to read a story like 'The Bloody Chamber' in crude, if reversed, Freudian terms, with the girl-child rescued from violent absorption in her lover/father by an avenging mother who, charging in on her horse like the cavalry, restores an alternative bourgeois feminist family order as mother and daughter live happily ever after, running a school. As ever, however, Carter reworks the categories: men do not act to guarantee the law but to break it, their passions emerging from an unholy realm of pre-symbolic excess. The Bluebeard villain who takes the daughter smells not of civilization but 'amniotic salinity',[91] and whilst his sexuality is visual, his effect on women is fragmenting, dismembering. In the kaleidoscope of bedroom mirrors our heroine becomes not the true adult wife but 'that multitude of girls I saw in the mirrors', and in intercourse 'a dozen husbands impaled a dozen brides while the mewing gulls swung on invisible trapezes in the empty air outside'.[92] By the final scene, however, the excess which here the man represents has been thoroughly repressed once more, and the child grows into adulthood in complicity with the mother. Men act to split women from themselves and plunge them back into the amniotic; mothers offer a route back (or forwards) to a precarious self-unity. Mothers are heroic and act rationally, fathers – the Bluebeard villains of Carter's parodic sexual nightmare – are fragmenting, bent only on the dismemberment of the subject. The mother must intervene to restore her daughter to civilization; the child grows into adulthood in complicity with the mother. However, much as Carter plays with this possibility, Elaine Jordan charges against decoding it in the manner of Bonaparte: 'you cannot lay a grid across her work and read off meanings from it, according to a law of the same'.[93]

Fevvers has no such Oedipal ancestry. Supplementing this long list of Oedipal daughters, in her penultimate novel Carter writes the tale of a fatherless child, hatched from an egg, and daughter only to her surrogate mothers. In *The Passion of New Eve*, a protracted parody of the undoing and redoing of utopian pre-Oedipal relations, a man is taken by an arche-

91 Carter, 'The Bloody Chamber', p. 12.

92 Ibid., p. 17.

93 Jordan continues: 'To offer one example . . . Patricia Duncker assumes that the blindness of the piano tuner in "The Bloody Chamber" is a symbolic castration, like Charlotte Bronte's mutilation of Mr Rochester in *Jane Eyre*' ('The Dangers of Angela Carter', in *New Feminist Discourses*, ed. Armstrong, p. 122).

typal 'Mother' and deprived of his manhood, turned into a woman, and turned out into the world. Like other tales of sex-change,[94] the process of adaptation to a new gender position is not easy; it is only when the 'new Eve' enters a cave-labyrinth at the close of the novel and in effect gives birth to herself, encountering the pre-Oedipal entirely internally, that she can emerge as a new woman, unparented and multi-parented, daughter and mother (as well as father) of herself. Finally, the identical-twin heroines of Carter's last novel, *Wise Children* – stars of vaudeville, musical comedy and cinema – grow up in a matriarchal household (in which even the grandfather clock is 'castrated'), but give birth to themselves as the fabrications of theatre entirely with reference to the mirrors of audience, image and their own reflections in each other.

Yet in her 1979 analysis of Sade's *Philosophy in the Boudoir*, which focuses on a daughter's (Eugénie's) sexual violation of her mother, Carter rethinks women's relationship to the Oedipal myth in quite different terms. Here she sets up Sade's wild argument *against* repression and *for* lawlessness by reading him through Freud:

> King Oedipus' transgressions were mother-incest and parricide; when he found out what he had done, he blinded himself, that is underwent symbolic castration. Eugénie, unlike Oedipus, acts in the knowlege she is committing a crime. Her crime is the culmination of her search for knowledge. She fucks her mother out of vengeance and so finds herself in the position of a female Oedipus but she is not blinded, she is enlightened.[95]

Monstrous as this is, it resonates with a larger experiment with representations of the mother taking place across Carter's work. Eugénie's act, concludes Carter, is against *the very fact of* her own existence; Sade's absolute atheism pushes his libertine to a position which 'cannot forgive the mother, not for what she is but for what she has done – for having thoughtlessly, needlessly inflicted life upon him'[96] (and her). Whilst Carter does not undo the processes of mothering in her work in quite such a radical way, nevertheless she fantasizes how they can be reimagined. Life in Carter is then 'inflicted' in a variety of ways: daughters are hatched from eggs, surgically constructed in caves; they even bring themselves into being in front of the mirror. Here I am thinking of Leilah's act of giving birth to herself as she makes up in the mirror of *The Passion of New Eve*, or the twin old women of *Wise Children*, making girls of themselves from the old images they have become: 'It took an age but we did it; we painted the faces that we always used to have on to the

94 Rose Tremain's marvellous 1992 novel *Sacred Country* explores the issue in a different, realist context.
95 Carter, *The Sadeian Woman* (London: Virago, 1979), p. 117.
96 Ibid., p. 135.

faces we have now.'[97] They have no children of their own, so they become their own children. These extraordinary images – alternative family romances – decentre the traditional image of the life-giver and nurturer, and reposition her somewhere 'in the distinction between woman and mother'. The mirror dramatizes yet another set of alternative identities, from which a final choice is never made. The play of making-up is kept open.

The fantasy of the subject's origins is, then, strung out across Carter's work in a protracted moment of hesitation. Fact or fiction? – What is she? – *Where* does she come from? These are questions Carter refuses to resolve, so that they remain unanswered. Her stories repeatedly return to the possibility of using visual fantasy to break down identity, or to expose the fact that the self is never finished or complete – what she writes are a series of visual challenges to the subject. If primal fantasy is the traumatic site upon which the self *begins* to be crystallized, Carter's work seeks to engage with that process at a later stage, in adult fantasies of infantile development, in family romances which hope to reimagine entirely the family and its fictions.

97 Carter, *Wise Children* (London: Chatto & Windus, 1991), p. 192.

4

Too Early and Too Late: Mrs Oliphant and the Wolf Man

> Analytic interpretation, it is said, not without some justification, is retrogressive: it dismantles present constructs in order to reconstitute anterior tendencies, wishes, sets of events belonging in the past.[1]

Freud's 'Wolf Man' case is uniquely concerned with deferral, deferred action, the time of the subject and the revision of the past in the present, which leaves neither past nor present the same. It concerns 'an infantile neurosis which was analysed not while it actually existed, but . . . fifteen years after its termination'.[2] From the start the analysis is set up as necessarily 'too late', after the scene, yet it is an analysis which is only possible because of this belatedness: it begins after what might seem to be the end, and proceeds retrospectively. This is psychoanalytic time-travel, this analysis which *looks back* toward the work which the subject has already done within itself when comprehending and revising its past in the construction and working-through of neuroses. In 'Analysis Terminable and Interminable' Freud writes of analysis 'undertak[ing] a *revision* of these old repressions';[3] but the revisions which take place in analysis are only working on an earlier set of revisions within the subject, known as deferred action. Of this, Jean Laplanche writes in the brief discussion of 'deferred action' in *New Foundations for Psychoanalysis,*

1 Jean Laplanche, 'Psychoanalysis, Time and Translation', in *Seduction, Translation, Drives* (London: ICA, 1992), p. 169.
2 'From the History of an Infantile Neurosis (the 'Wolf Man')' 'Wolf Man', PF vol. ix, p. 235.
3 Freud, 'Analysis Terminable and Interminable', p. 227; my italics.

> Analytic observers are fated to being either too early or too late, not because some ill-defined metaphysical curse has been laid upon them, but because their object itself is constructed in two stages.[4]

The last chapter involved a discussion of Kleinian child analysis, which aims to treat the patient as early as possible in order to fend off this analytic belatedness. This early moment which is Klein's focus has had some importance for cultural analysis, as we have seen, particularly in work which traces the pre-Oedipal and the role of the mother in texts. This chapter moves forward in the life of the analysand, and back to Freud again, with reference to Laplanche's work on deferred action, to think further about how the subject's sense of its past offers a unique intersection with literary and cinematic thinking on the self in time. If subjects come into being through their relationship with narratives, then narratives are formed in time; but, as we shall see, the form of narrative time under analysis here does not flow in only one direction.

At the opening of the 'Wolf Man' case Freud argues against the effectiveness of child-analysis on the grounds that 'too many words and thoughts have to be lent to the child' by the analyst.[5] Instead he sets up the 'analysis of a childhood disorder through the medium of recollection in an intellectually mature adult'. The key displacement of this text is already clear: if the Schreber case is marked by a spatial as well as a literary-critical displacement (Freud never met Schreber, the subject of that analysis, but instead analyses him through reading the written text of Schreber's *Memoirs*), the 'Wolf Man' case is predicated on temporal displacement, the analysis of a past event involving its inevitable revision and rewriting in the present. Analysis based on recollection thus, 'necessitates our taking into account the distortion and refurbishing to which a person's own past is subjected when it is looked back upon from a later period'.[6]

'Distortion and refurbishing', if presented rather negatively here, could also be said to be positive qualities in the literary critical process, as well as in the process through which the past is reworked and read at a later moment in a variety of narrative-based cultural forms. I want to do several things in this chapter, and one of them is to move between Freud's 'Wolf Man' case and Margaret Oliphant's 1899 *Autobiography*, via some interjections by Laplanche on deferred action or 'afterwardsness' (*nachträglichkeit*). It might be said that Victorian narrative prose, fictive or biographical, is uniquely concerned with the time of the self and its impact on narrative form. The 'classic' characterizations of the Victorian

4 *New Foundations for Psychoanalysis* (Oxford: Blackwell, 1989), p. 88.

5 '[E]ven so,' he continues, 'the deepest strata may turn out to be impenetrable to consciousness' (*Wolf Man*, p. 235).

6 Ibid., p. 325.

novel reveal not purposeful existential souls, morally wilful and in control of their monolithic fates, but subjects divided by their involuntary inhabitating of a past they cannot shake off. Smitten by a belatedness which ought to be modern, the Victorian subject is characterized by a feeling out-of-place. She is haunted by the presence or the loss of the past, living out of time, alienated from her origin and a half-remembered past which is as tenacious in its grip on the self as it is obscure. Something in the past of the self cannot be done with. The classic narrative structure of the novel might appear to be governed by an unbending teleology, guiding the realist plot from the alpha of the opening sentence to the omega of the final full stop, but a stronger countercurrent drags the substance of the text *back*, through guilt, hauntings, the anxiety of lost origins, knots of early trauma and neurosis, which mean that the imagined past of Victorian prose is seldom properly closed. In particular, Mrs Oliphant's *Autobiography* is written so as to suggest that it is essentially *anterior* to itself.

I am not suggesting that characters behave on the page as the 'Wolf Man' does in Freud's case, 'real people' trapped by a fictive neurosis which the psychoanalytic critic, in conjunction with the novelist, can help them shake off. What I want to work through here is the possibility that we can understand something more of the double movements of time and confused sequence in the texts themselves, which look back as much as they look forward in a pattern of traumatic cross-currents, by relating them to these thoughts on the 'afterwardsness' of the psychoanalytic subject, and the way in which psychoanalytic narratives move backwards and forwards in time. In this chapter I shall also look at John Boorman's cinematic essay on masculinity, collective repression and flawed forgetting, the 1972 film *Deliverance*, both because of its work on the narrative past which is resolved in different ways, and because it offers a modern, visual realization of some of the curious time-processes at work in Oliphant, which also implicate both audience and reader in their curious temporality.

I
Nachträglichkeit and the Wolf Man

In a letter to Wilhelm Fliess written on 6 December 1896, Freud opens up a new angle on the textual life of the analytic subject by describing the way in which memory is rewritten through its repetitions in time:

> I am working on the assumption that our psychic mechanism has come into being by a process of stratification: the material present in the form of memory

> traces being subjected from time to time to a *rearrangement* in accordance with fresh circumstances – to a *retranscription*. Thus what is essentially new about my theory is the thesis that memory is present not once but several times over.[7]

The notion of deferred action has been most fruitfully explored by Laplanche, first with Pontalis in 'Fantasy and the Origins of Sexuality' and the entry on 'Deferred Action' in *The Language of Psycho-Analysis*. In addition, two short texts by Laplanche writing alone (which are now available in English) also discuss the time of the subject: 'Notes on Afterwardsness', and 'Psychoanalysis, Time and Translation'. These texts each open up a notion of retrospective experience which itself lies at the heart of the connection between psychoanalysis and other forms of narrative, in the practice of rewriting a past event in the present, or rather the emergence of the significance of the past event occurring at a later moment ('memory is present not once but several times over'). Laplanche and Pontalis trace the history of the term *nachträglichkeit* ('afterwardsness' or 'deferred action') as a term covering

> experiences, impressions and memory-traces [which] may be revised at a later date to fit in with fresh experiences or with the attainment of a new stage of development.[8]

Deferred action for Freud is, then, not the Jungian notion of retrospective phantasies ('the adult reinterprets his past in his phantasies . . . [as] a way for the subject to escape from the present "demands of reality"'[9]), nor is it the existential notion of revision of the past in conformity with a conscious project. Rather, what is 'revised' is an unassimilated or traumatic sexuality, 'whatever it has been impossible in the first instance to incorporate fully into a meaningful context'.[10]

Laplanche's other work is also saturated with the concerns of *nachträglichkeit*, in a way which informs his understanding of individual history and the history of psychoanalysis. At the opening of *New Foundations* he writes:

> Psychoanalysis shows us that history is neither a continuous nor a cumulative process, that it does not have a happy ending, that it does not evolve smoothly and that its course is marked by repression, repetition and the return of the repressed.[11]

7 Freud, *Complete Letters to Wilhelm Fliess*, p. 207; Freud's emphasis. Part of this is quoted by Laplanche and Pontalis in *The Language of Psycho-Analysis*.
8 Laplanche and Pontalis, *The Language of Psycho-Analysis*, p. 111.
9 Ibid., p. 112.
10 Ibid.
11 p. 2.

Nachträglichkeit articulates this most acutely, but it is also present at other moments in Laplanche and in the very way in which he situates himself in relation to Freud. It is not only individual history which is 'neither a continuous nor a cumulative process', but the history of psychoanalysis itself. The title of *New Foundations for Psychoanalysis* implies a tension or an interplay, between the terms 'new' and 'foundations': we are going back to our foundations in order to renew them'.[12] Laplanche's is not a 'return to Freud' (like Lacan's famous move): 'I would prefer to speak of *going back over* Freud, as it is impossible to return to Freud without working on him, without making him the object of work.'[13] This renewal through 'going back' is also the project of the fiction and film I shall turn to shortly, since this is also what cultural texts do: rather than 'return[ing] to' the problem, they work over it and put it to work.

The sense of history evoked here is of a process which continually undoes and rewrites the past. Equally, the time and history of the subject is directed *back* as much as it is directed *forwards*. Repression, repetition, the return of the repressed, are all marked by an attitude to time which severely disrupts the linear model which posits it as moving forward in a smooth teleological progression: for psychoanalysis, the past is never entirely past – nothing is forgotten. Freud discusses the 'timelessness' of the unconscious in a number of places. In the important metapsychological paper of 1915, 'The Unconscious', he writes:

> The processes of the system *Ucs.* [the unconscious] are *timeless;* i.e. they are not ordered temporally, are not altered by the passage of time; they have no reference to time at all. Reference to time is bound up . . . with the work of the system Cs [consciousness or the pre-conscious].[14]

Time only comes into operation when unconscious phenomena are brought through into consciousness (or the pre-conscious) through the process of analysis itself. Freud also meditates on this in the 'Wolf Man' case (roughly contemporaneous with 'The Unconscious'), where he charges the physician to behave 'as 'timelessly' as the unconscious itself'.[15] In analysis the subject negotiates the connection between the timelessness of the unconscious and the temporal nature of her conscious narrative. Moments in her history may be temporally separate in terms of how they are perceived consciously, but they are intimately bound together in the way in which the unconscious works upon them. However, as Peter Brooks argues in 'Freud's Masterplot', the literary nature of narrative means that its sequential, 'only-in-one-direction' aspect is also a trick. In that narratives rely on repetition – a going-back

12 p. 3.
13 Ibid., p. 16; his emphasis.
14 PF vol. xi, p. 191.
15 *Wolf Man*, p. 237.

(retrospection) as well as a looking-forward (anticipation), they have the power to slip out of and between times in a way akin to Freud's conception of the unconscious. Brooks writes of repetition and narrative time:

> Narrative must ever present itself as a repetition of events that have already happened, and within this postulate of a generalized repetition it must make use of specific, perceptible repetitions in order to create plot, that is, to show us a significant interconnection of events. Repetition is a *return* in the text, a doubling back. We cannot say whether this return is a return *to* or a return *of*: for instance, a return to its origins or a return of the repressed. Repetition through this ambiguity appears to suspend temporal process, or rather, to subject it to an indeterminate shuttling or oscillation which binds different moments together as a middle which might turn forward or back.[16]

How does this operate in the *Wolf Man* case, in a way which in itself constitutes a kind of literary process? If analysis is either 'too early or too late', this means that the 'Wolf Man' (both the case and the man) is never really present or in the present. Freud and his analysand look back, and the moment they look back *to* is itself destined to be left unresolved until the future, as a loop of trauma is set up. The child is the site of a neurosis which bears fruit in the adult, and the adult looks back (analytically) to himself-as-child, completing the loop in order to break it. But the deferral or sequence of deferrals and deferred understandings upon which the case rests has itself already taken place by the time Freud and the 'Wolf Man' meet, for the moment the analysand looks back *to* is itself (at least) double, and buried in early childhood.

The young man who is to be known as the 'Wolf Man' (from his early anxious dream and phobia of wolves[17]) comes to Freud at the age of eighteen, but it is his early problems which are the focus of the case: 'an anxiety-hysteria (in the shape of an animal-phobia), [which then changes] into an obsessional neurosis with a religious content'.[18] The case hinges on a lengthy analysis of the feared wolf-dream, itself simple and undramatic – in its surface-narrative at least. At three or four years old, the analysand has a dream in which he is asleep in his own bed. The window in his bedroom opens, and sitting motionless in the walnut tree outside the window are six or seven white wolves, who simply stare at him calmly:

> the wolves sat quite still and without making any movement on the branches of the tree, to the right and left of the trunk, and looked at me. It seemed as

16 In *Literature and Psychoanalysis: The Question of Reading: Otherwise*, ed. Felman, pp. 280–300; quotation p. 288.

17 I shall refer to the analysand himself as the 'Wolf Man' and the case itself as *Wolf Man*.

18 *Wolf Man*, p. 234.

> though they had riveted their whole attention on me. – I think this was my first anxiety-dream.[19]

A simple scenario, which Freud manages to read through a system of reversals back to the analysand's memory of the primal scene. This 'tracing back' consequently happens as the case proceeds *forwards*.

Unravalled sequentially, this is how things run: first it is posited that at eighteen months the 'Wolf Man' witnesses (or, as is later argued, fantasizes[20]) a scene of sex between his parents, involving intercourse from behind, in the position of animals. But it is only at the age of four, with this dream, that 'a deferred revision of the impressions so received . . . penetrate the understanding'.[21] The word 'penetrate' here implies that the boy is *made* to understand the significance of the sex scene, passively, by a realization coming later from the outside, but the understanding, the penetration and the revision are all his, and he does the very *active* work involved in the deferred process of revision. At one point Freud also calls this moment of the dream a 'revival' of the primal scene[22] which produces an 'alteration' in the subject. He then avoids using the passive word 'recollection' to refer to what happens to the boy at four (putting it under erasure by *stating* his avoidance of it[23]), instead arguing that it was at this stage that the scene was 'activated', and that it is this traumatic 'work' which is important.

So the dream at four gives the boy 'a deferred comprehension of the scene he had experienced when he was one and a half';[24] but how exactly do the wolves stand in for the primal scene? In short, the dream itself presents a reworked version of the earlier scene, which Freud reads through the reversals it presents. The wolves are motionless, and they stare at him calmly, both of which are interpreted through the distortion, 'transposition or reversal' which the original scene has undergone in the dream-work, the boy's unconscious processing of it into the dream. The dream is then the primal scene in reverse. The child *looks at* the scene of parental sex, and this is distorted so that he *is looked at* in the dream. If the

19 Ibid., p. 259.
20 The scene may be something of both: '[I]t must have been a *coitus a tergo, more ferarum*' Freud coyly writes. But he continues: 'Perhaps what the child observed was not copulation between his parents but copulation between animals, which he then displaced onto his parents, as though he had inferred that his parents did things in the same way' (*Wolf Man*, p. 292).
21 Ibid., p. 270.
22 Ibid., p. 275.
23 What I mean by this is that Freud *tells us* that he is not using the word – 'the activation of this scene (I purposely avoid the word "recollection")' – in a way which still allows the possibility of 'recollection' to hover over what he is saying. Something is recalled *as* it is activated (*Wolf Man*, p. 276).
24 Ibid., p. 315.

original scene was one of 'the most violent motion', in the dream 'the wolves sat there motionless; they looked at him, but did not move'.[25]

The boy has then retold the story to himself through reversal in the dream-work, partly as a way of taking control, partly as a process of displacement. Later, Freud writes that it was about the time of the boy's fourth birthday 'that the dream brought into deferred operation his observation of intercourse at the age of one and a half . . . The activation of the picture, which, thanks to the advance in his intellectual development, he was now able to understand, operated not only like a fresh event, but like a new trauma.'[26] Further still, the scene of parental sex is itself interpreted as a repetition of an earlier event, in which the boy's sister 'seduces' him by playing with his penis. Each event acts upon and revises those before:

> the sexual development of the case . . . was first decisively influenced by the seduction, and was then diverted by the scene of observation of the coitus, which in its deferred action operated like a second seduction.[27]

'[I]t always takes two traumas to make a trauma,' writes Laplanche, arguing for a psychoanalysis which can read the significance of those two moments of trauma *together*: 'deferred action is a two-stage mechanism, and neither of its stages can be detected on its own.' As Freud's work moves on, so the possibilities for revisions and deferred understanding change; in *The Language of Psycho-Analysis* Laplanche and Pontalis point out that for the Freud of the Seduction Theory, such belated comprehensions take place at puberty: 'only the occurrence of the second scene [awakening of sexuality at puberty] can endow the first one [seduction/primal scene] with pathogenic force.'[28] In their discussion of the Seduction Theory in 'Fantasy and the Origins of Sexuality' this is opened up further: human sexuality is characteristically 'temporal', and trauma takes place in 'two stages', the second of which 'can be conceived only as arising from something *already there*, the reminiscence of the first scene'.[29] Later Freud sticks to this basic idea of psychic rereading, only shifting the moment of revision back from puberty to early childhood, as we can see from the 'Wolf Man' case.

Deferred comprehension is, then, a *reading* of the scene, an interpretation which generates its traumatically charged meaning, enacted by the boy unconsciously and constructed at a distance. Something is set in motion at one and a half for the 'Wolf Man', but it gets under way *again* as 'deferred revision' only later; '[t]he effects of the scene were deferred,'

25 All quotations relating to this discussion, ibid., p. 266.
26 Ibid., p. 351.
27 Ibid., p. 280.
28 Laplanche and Pontalis, *The Language of Psycho-Analysis*, p. 113.
29 Laplanche and Pontalis, 'Fantasy and the Origins of Sexuality', p. 5; their italics.

Freud writes, 'but meanwhile it had lost none of its freshness in the interval between the ages of one and a half and four years'.[30] Thus the 'Wolf Man' only begins to understand the significance of the scene

> at the time of the dream when he was four years old, not at the time of the observation. He received the impressions when he was one and a half; his understanding of them was deferred, but became possible at the time of the dreaming to his development, his sexual excitations and his sexual researches.[31]

Yet later in the case Freud argues that it is the boy's view of dogs copulating, a spectacle seen only months before his awful dream, which is superimposed onto a *fantasy* of his parents. This has the effect of shrinking the key period of deferral from that between eighteen months and four years (from 'actual' primal scene to dream) to just a few months during his fourth year, in which the spectacle of the copulating dogs is superimposed onto an 'innocent' image of his parents, culminating in the dream. This, then, is one moment of revision, a revision which takes place in the case history itself as a written text, as well as in the child's fourth year:

> What supervened during the expectant excitement of the night of his dream was the transference on to his parents of his recently acquired memory-picture [of the dogs] . . . He now arrived at a deferred understanding of the impressions which he may have received a few weeks or months earlier. . . .
>
> . . . It is at once obvious how greatly the demands on our credulity are reduced. The period of time during which the effects were deferred is very greatly diminished; it now covers only a few months of the child's fourth year and does not stretch back at all into the first dark years of childhood.[32]

As with the cases discussed in Chapter 1, this case also builds up like a short story or a work of detective fiction: just what is deferred in the 'Wolf Man's' early life, and for how long, is itself suspensefully withheld within the narrative of the case as a whole and as a story. Like *Dora*, this is also a fragmented text: 'I have abstained from writing a complete history of [the] illness,' writes Freud, focusing only on its infantile phase, so we never arrive at the 'present' moment at which our narrator is presumed to sit in full and realized knowledge of what he has told us.[33] Freud is nothing if not adept at spinning the tale of his case by withholding, or deferring, its dénouement. He stops short only to go back, and sometimes he reads backwards, keeping earlier observations on hold: 'We must here break off the discussion of [the 'Wolf Man's'] sexual

30 *Wolf Man*, p. 277.
31 Quoted by Laplanche and Pontalis, *The Language of Psycho-Analysis*, p. 113.
32 *Wolf Man*, pp. 292–3.
33 Ibid., p. 234.

development until new light is thrown from the later stages of his history upon these earlier ones,'[34] he writes, and in the dream analysis he works in reverse: 'Let us take this last remark as a starting-point.'[35] The case is peppered with promises which also distance us from what is to come: 'I *shall* begin'; 'infantile neurosis *will be* the subject of my communication', he writes, of 'the analysis which *is to be* described in these pages'.[36]

This deferral on the level of rhetoric is partly an effect of the deferring nature of the illness itself which Freud is analysing (content has then infected form), but it is also necessitated by the conventions of narrative, since narratives depend upon deferral. Suspense, what Peter Brooks has called 'Reading for the Plot', relies as much upon witholding information as it does upon releasing it, saying too little in the right place, encouraging a process of gradual connotation on the part of the reader so that events placed early in the story make more sense as the reading proceeds. In this way links are made, connections and sequences established, the reader *looks back* as she *reads on*. Reading (perhaps *all* reading) proceeds in two different directions at once. This is partly about what Brooks calls textual 'forepleasure', which suggests,

> a whole rhetoric of advance towards and retreat from the goal or the end, a formal zone of play (I take it that forepleasure somehow implicates foreplay) that is both harnessed to the end and yet autonomous and capable of deviations and recursive movements.[37]

Before pursuing this further, I also want briefly to situate the case as a key site for the analysis of the interpenetration of subjectivity and the actual stories which circulate in a culture. In the central section of the case, Freud quotes himself, reproducing five pages verbatim from the 1913 essay, 'The Occurrence in Dreams of Material from Fairy Tales',[38] in which he had already published his analysis of the dream under discussion. This is, then, a text displaced from its 'original' place, its meanings broadened and revised by the new context in which it finds itself. Yet it is first published in a way which suggests that the 'real' place in which the dream is to find its true context will come later, with the publication of the full case. There is also therefore a deferral at work in Freud's publishing history – he gives us something of the analysis in 1913, but we have

34 Ibid., p. 280.
35 Ibid., p. 264.
36 Ibid., pp. 240, 234, and 239. These examples come early on in the case, but even as it is drawing to a close the rhetorical device is maintained: '*If* we look at the matter more closely we *shall* see that . . .'; 'We *must now* summon up our courage . . .' (p. 356); and 'I *will* conclude my survey' (p. 361), four pages from the end. All italics mine.
37 'The Idea of a Psychoanalytic Literary Criticism', in *Discourse in Psychoanalysis and Literature*, ed. Shlomith Rimmon-Kenan (London and New York: Methuen, 1987), pp. 1–18, quotation p. 7.
38 *SE*, vol. xii.

to wait until 1918 for the case as a whole to be published – the 1913 section is peppered with further rhetorical deferrals and promises: 'I shall have . . . to deal with this dream again elsewhere', 'I shall discuss this on another occasion', he writes, but when the full case comes it turns out to be also a narrative fragement, not the whole story at all – Freud 'abstains' from writing a complete history of the illness.

Texts within the text of the *Wolf Man* abound, operating rather like the structure of a (usually modernist) novel deploying a variety of narrators, telling stories in a 'Chinese-box structure' of layers.[39] The 'Wolf Man's' texts – his stories of the primal scene, the dream, the seduction – also continue to be revised as their place in the narrative of the case changes. Whilst Freud seems to be offering a meta-narrative, a gloss on the stories within his story, his wider role as *unreliable* narrator (see my earlier discussion of *Dora*) problematizes this, as does the fact that self-quotation simply presents Freud's own text as one amongst others. Perhaps more important than this, however, is the issue of the influence of the literary upon the unconscious, the interweaving of elements of both into each other, which the case also raises, since it is an experience of fiction which activates and provides the imagistic co-ordinates upon which the boy's anxiety is fixed. Freud's self-quotation comes from a text which shows that the dream of the wolves is woven from fragments of fables, which are then used by the subject as instruments for reading his own individual history. Story-book pictures become illuminations of the 'Wolf Man's' tale; indeed, in one instance fictional images directly illustrate and provoke his trauma, when his sister torments him with a representation of a wolf from a picture-book.[40] In particular, 'Little Red Riding Hood', 'The Wolf and the Seven Little Goats' and a story from *Reynard the Fox* are all put to work in providing the imagery for the child's anxious personal narrative.[41] This is an intertextuality which operates temporally, as different time-zones interpenetrate the 'timelessness' of fable with the specific unfolding of personal history. The displacement of

[39] Nancy Armstrong uses this term in relation to the complex narrative structure of *Wuthering Heights*, in 'Imperialist Nostalgia and *Wuthering Heights*', collected in *Wuthering Heights* by Emily Brontë, ed. Linda H. Peterson (Boston: St Martin's, 1992), p. 430. Other essays collected in this edition discuss the novel's narrative structure, but of primary interest in this context is Philip K. Wion's essay, 'The Absent Mother in *Wuthering Heights*', and J. Hillis Miller's '*Wuthering Heights*: Repetition and the "Uncanny"'.

[40] The 'Wolf Man' becomes directly involved in the fictional representation of the image: 'Whenever he caught sight of this picture he began to scream like a lunatic that he was afraid of the wolf coming and eating him up. His sister, however, always succeeded in arranging so that he was obliged to see this picture, and was delighted at his terror' (p. 243).

[41] Their 'unreality' is their prime characteristic, when set against the awful lasting reality of the dream: 'The dream seemed to point to an occurrence the reality of which was very strongly emphasized as being in marked contrast to the unreality of the fairy tales . . . behind the content of the dream there lay some such unknown scene' (*Wolf Man*, p. 264).

images and *vignettes* from literary history (albeit the fairy-tale) thus acts upon the case's sense of sequence; a story heard at one point in the child's life lends its imagery to the meanings and interpretations being constructed at another point. Thus the subject's narrative not only works against a linear sense of narrative order; it also comes from a number of sources.

This is only one sense in which the *Wolf Man* is an exemplary case for an understanding of narrative, of how the narrative of stories, subjective experiences and psychoanalytic processes are all informed by a 'literary' understanding of time. It is not simply through its emphasis on unconscious word-play, punning or linguistic patterning, or even its reliance on literary myth (Oedipus or fairy-tales), which makes the way in which psychoanalysis understands the subject as essentially 'literary'. Just as important (but rather more neglected by critics) is the way in which psychoanalysis reads the subject through a model of narrative. The centrality of deferred action or 'afterwardsness' to the psychoanalytic process underpins the fact that the 'literary', involving these acts of rereading and revision as well as the need to articulate experience in narrative, is something the subject inhabits at a fundamental level.

II
Masculinity and Collective Disavowal in *Deliverance*

> *Nachträglichkeit* means many things in German; it means the fact that something comes afterwards; it means also something like revenge or bearing a grudge, when you take revenge because of something that took place before.[42]

Many psychoanalytic experiences resonate with a curious temporality. The hysteric, as we have seen, 'suffers from reminiscences'. Tania Modleski picks up on this in her account of melodramatic cinema as an hysterical form: just as, for the hysterical subject, what is repressed returns to and through the body as symptom, so 'in melodrama, what is repressed at the level of the story often returns through the music or the mise-en-scène'.[43] *Mise-en-scène* is thus read as the film's body, the site of symptomatic signs given out visually which the film's narrative cannot bring itself to express. Modleski takes this further, by arguing that melodrama is hysterical in its inability to move on, its incessant need to look back, its confusion of past and present:

42 Jean Laplanche, *Seductions, Translations, Drives*, trans. Martin Stanton (London: ICA, 1992), p. 42.
43 'Time and Desire in the Woman's Film', in *Film Theory and Criticism*, ed. Mast *et al.*, p. 537.

> Unlike most Hollywood narratives, which give the impression of a progressive movement toward an end that is significantly different from the beginning, much melodrama gives the impression of a ceaseless returning to a prior state. Perhaps the effect may be compared to sitting in a train watching the world move by, and each time you reach a destination, you discover that it is the place you never really left. In this respect melodrama appears to be quite closely linked to an hysterical experience of time and place. The hysteric, in Freud's formulation, suffers from reminiscences. In melodrama, the important moments of the narrative are often felt as eruptions of involuntary memory, to the point where sometimes the only major events are repetitions of former ones.[44]

I want to distinguish the hysterical 'impression of a ceaseless returning to a prior state' from the phenomenon of deferred action which is under discussion in this chapter.[45] The hysteric is haunted by the past in terms of its eruption into the present as bodily symptom, and the inability to escape a past repressed but not erased which is reactivated at puberty. What distinguishes deferred action from this 'suffering from reminiscences' is the process of rereading, of analysis and resignification, which takes place second time around, so that the first event is only really activated through the second. Deferred action is not a 'delaying discharge'; Laplanche and Pontalis write that it 'cannot be understood in terms of a variable time-lapse, due to some kind of storing procedure, between stimuli and response';[46] there is work involved in the second stage – 'a "work of recollection" which is not the mere discharge of accumulated tension but a complex set of psychological operations'. The eruption of the past into the present in certain forms of hysteria is then quite a different prospect from the process through which an early experience is reworked later. Again, for reasons of argument, *Cousin Phillis* could provide an example of the 'time-bomb' theory of development. As we saw in Chapter 1, Phillis's is a crisis which is just *waiting* to happen, and its blossoming in the body language which exudes from her silence is the next step of a process which is essentially linear. The seed of discontent is sown at one moment in the narrative, and so it grows and blooms in another. For Laplanche, however, '"afterwards" is not the after-effect but it is also *something going back to before*'.[47] Phillis's hysteria is not about going back to before (though perhaps Paul's unconsciously self-analytic narration *is*); it is about sinking deeper, confirming the terror of the original thought, rather than rethinking it, and only really understanding it, at a later moment. She does not reread or rework her betrayal in the second

44 Ibid., p. 540.

45 This is also to be distinguished from the compulsive repetition which I shall discuss in Chapter 5, which, as we shall see then, is most clearly identified by Freud in terms of the death drive.

46 Laplanche and Pontalis, *The Language of Psycho-Analysis*, p. 114.

47 Laplanche, *Seductions, Translation, Drives*, p. 41; my italics.

step of her illness; rather, the illness is the logical (or illogical) conclusion of the betrayal. For Phillis, then, it only takes *one* trauma to make a trauma.

However, something else is operating in the narrative of John Boorman's 1972 film *Deliverance*, and I want to pause on this in order to set up some other possibilities for examining how psychoanalysis might address the subject's relationship to a traumatic past. Even in its title, the film suggests a difficulty in shaking something off. According to the *Shorter Oxford English Dictionary*, 'deliverance' is both 'The act of setting free, or fact of being set free; liberation, release, rescue' and 'The action of giving up; surrender', a double-edged meaning, of liberation and submission, active release and passive acquiescence to bondage, which characterizes what I wish to investigate in the uneasy dynamic of this film. Four city men from Atlanta set out to canoe down 'the last wild, untamed, unpolluted, unfucked-up river in the south', led by macho pseudo-philosopher Lewis (Burt Reynolds), whose proclamations on survival in the wilderness narrate us through the first part of the film. This is a drowning wilderness: the monolithic 'power company' we see bulldozing and dynamiting as the credits roll, moving churches and digging up coffins as the film draws to a close, are engaged in a wholesale act of ecological 'rape' (as Lewis puts it), damming the river and flooding everything in its wake. The feeling that something is in the process of being lost or buried, that this is a 'last chance', pervades the film. Indeed, the poor white trash backwoods folk with whom Lewis and friends soon clash are themselves on the brink of displacement, if not extinction. This sense of loss soon becomes an active need on the part of the protagonists to *lose*, to forget, to repress the traumas they experience as the film unfolds. A central dynamic is what is at stake in how we try to give up the past, and more importantly how we *fail* to do this, in the face of the insistent return into the present of past events.

The trip unravels into an antediluvian nightmare of pursuit, as the sinister, 'genetically deficient' mountain men hunt our heroes down-river: in the most infamous scene, Bobby (Ned Beatty) is anally raped as Ed (Jon Voight) is forced to look on, only to escape oral rape himself when Lewis kills Bobby's rapist. Drew (Ronny Cox) then dies in ambiguous circumstances, reappearing a little later like the repressed returning, as a horribly twisted corpse. The men conjecture that Drew was shot by Ed's would-be-rapist. Lewis is badly injured and relinquishes control to Ed, who shoots the man they *think* is Drew's killer, and the attempted rapist of Ed. The three survivors escape to an unnerving safety built upon a cover-up story which disavows not only their double murder but – more importantly – the shame of male rape and the uneasy circumstances of Drew's death.

This brief précis suggests what it is that the men must unsuccessfully forget, or rather *fail* to forget and then be forced to *re-encounter* despite their efforts at collective repression. The film offers a succinct focus for the ways in which a traumatic past is reworked (unsuccessfully) at a later moment – rewritten, in fact, so that things we remember witnessing (as an audience) are half-undone in the character's anxious discourse, leaving an unnerving sense that we are watching a film whose past is constantly subject to revision in the 'present'. Before elaborating on how this is worked through, I need to say a few more things about what it is in particular which makes this film so interesting to a psychoanalytic reader.

Firstly, at the heart of the film are the 'mountain men', perhaps its most unnerving element. The woods they inhabit are 'real deep – inaccessible' – the fantasy *mise-en-scène* of a deeply transgressive Grimm's fairy-tale. Somehow too close to nature, the men's identities are marked and muddied by being 'too natural' to be properly human (the 'mountain' prefix qualifies their status as 'men'), too intimate with, and isolated by, the natural (American) landscape to be wholesomely American ('There are some people up there that ain't never seen a town before' says Lewis). So natural are these people that they have become unnatural (*too* natural), but the precise nature of their 'naturalness' is characterized not by an idealized authenticity but by something darker. Critics often note the cinematic conventions governing the representation of Deep South, down-home country folk – bad teeth, worn dungarees, rusty cars, lazy demeanour appropriate to long hot afternoons lounging on the peeling verandah – as well as their willingness to go one step further as the monstrous hillbillies of horror.[48] Bobby's opening observation about the detritus of civilization washed up at their first stop – 'Look at the junk. I think this is where everything finishes up. We may just be at the end of the line' – is already laced with threat. The sense that here, too, nothing (and everything) is lost underscores the suggestion that they have arrived in the heart of the American unconscious, a 1970s cinematic refiguration of Conrad's Congo in *Heart of Darkness.*

The 'horror of incest' which Freud discusses in a central section of *Totem and Taboo* clearly has little power here. Country folk are quite simply interbred, bearing the visible cinematic signs of their natural unnaturalness in a way which we have learned to read clearly. The rural family

48 The horror potential of a country-versus-city clash, charged up by varying degrees and histories of economic exploitation, cultural misunderstanding and sexual threat, has been a rich cinematic resource in the last thirty years or so, well-charted by Kim Newman, Carol J. Clover and others. *Straw Dogs* (1971), *The Hills Have Eyes* (1977), *Southern Comfort* (1981), the *Texas Chain Saw Massacre* films (1974 onwards) (to name only a few) all play with a similar bank of city-meets- (and defeats) - country elements. The family too has long been a staple source for horror, as breeding-site for psychosis, indulged taboo desires, mutant or demonic children, mostly being cinematically born from the late 1960s through into the mid-1970s into the families of urban (and often prosperous) America – *Rosemary's Baby* (1968), *The Exorcist* (1973), *The Omen* (1976), *The Fury* (1978), *Carrie* (1976), even *Eraserhead* (1977).

of cinema horror is disturbing for classically Freudian reasons: this is a composite monster rendered from recognizable elements of backwoods Americana and the awful suspicion that interbreeding and literal mother-fucking have shredded the vestiges of behavioural control. These are people who might just do *anything*, because they have already done, or been begotten by, the worst possible thing. Uneasy glimpses of bodies which (we assume) wear the symptoms of incest push the film into difficult territory from the start. 'All these people are related,' Lewis says after dispatching the rapist, their interconnection entirely undermining the law: 'I'll be goddamned if I'm gonna come back up here and stand trial with this man's aunt and his uncle, maybe his mom and his daddy sitting in the jury-box.' We already know something of this from the half-glimpsed handicapped child watched over by his grandmother, and the visible difference of the albino and strangely androgynous retarded boy with the banjo, striking out the famous 'duelling banjos' theme with uncanny dexterity whilst also challenging us (along with the Atlantan invaders) to make the judgement of subhumanity. 'Talk about genetic deficiencies,' says Bobby. 'Ain't that pitiful?'

So rather than showing the consequences of the death or the absence of the family, here the city men become entangled in a sticky genetically intensified web made by *too much* family. If the family done right (for psychoanalysis, at least) ideally ensures and facilitates the smooth path into cultural conformity, the family done wrong breaks open the membrane of civilization with alarming (and here visually evident) consequences. The unnatural disturbances to human 'normality' brought about by an all-too-natural ignorance or defiance of the incest taboo (a defiance which, it seems, all country people live, breathe and breed by) is at work here. Perhaps it is this which underpins all that is most unnerving on screen and in implication – an overwhelming sense of relatedness run wild.

This anxiety about the family also underpins an anxiety about sexual identity, which also has implications for the way in which we shall think about time and the subject's past in a moment. *Deliverance* is often read as a masculinity-in-crisis film, within which these four men, more or less emasculated by city life, have to recover their lost (but not dead) 'Iron John' manhood. However, this is not so much a film about men looking for an authentic gender-identity as a film about masculinity as an agent of change and difference, something which, far from being fixed, immutable, sovereign, can be lost and precariously found, diminished, or warped. Beyond the surface conflicts of 'men versus the elements' or 'civilized masculinity versus the rednecks', the film is caught up in a protracted encounter between men and their male 'others'. This is a 'man's film' in more ways than one (just as it shows that to be a man is rather more than one thing). Whilst family abnormalities underpin all that is wrong here, these are families who keep their women in the margins.

Women only creep around the edges and are always seen indoors: from our brief glimpse, ten minutes into the film, of a grandmother seen through a window, to the last fifteen minutes, when Ed meets a nurse and Bobby discusses large cucumbers with another old woman, the only 'real' woman's face we see is, fleetingly, a photograph of Ed's wife, which slips from his fingers as he climbs the cliff-face.

As the country family closes ranks, the city family begins to crumble. The woman in the photograph stands with her small son, and both of them are lost when Ed drops the photo. In the 'funeral' speech Ed gives for Drew, the family is evoked but in a way which underscores the fact that Drew has failed them. Ed stands in the water, embracing Drew's body as he speaks, staring into nothing, but also curiously staring *back* to a prior moment of safety, when normal familial relations and sure sexual identity prevailed. Repeated cuts between Ed embracing the supine Drew and the mute Lewis, now injured, lying in a similar pose in the canoe, suggest that Ed might as well be burying them all. In the middle of all this, one of the few bright objects is Ed's wedding-ring on the hand which holds the rock to which Drew is tied. But just as women and families are evoked as comfort and justification, so they are lost. Ed's speech underlines the uneasy distinction between presence and absence through its shift from from third person to first: 'He was a good husband to his wife Linda, and – uh – you were a wonderful father to your boys, Drew, Jimmy and Billy-Ray, and if we come through this I promise to do all I can for them.' He is talking to, and looking at, Drew, himself, and nothing – this is as much about the loss of his own family as it is about Drew's family's loss of their father. However, although women are almost entirely absent, this does not mean that the film is populated by only one sex (or, of course, that only men will find it interesting). It is not merely that there are no women here which suggests that men become less than or more than men in certain circumstances, but that marks of sexual identity and difference are set going in a process of mutation, intensified by the incestuous sexual pressure-cooker of the world of the film. Rather than guaranteeing stability or the law here, men only accelerate its breakdown.

Alongside the conventional narrative of a fairly classic moral-crisis film (which argues that heroes need to adopt the tactics of monsters in order to defeat monsters, thus becoming monstrous themselves) is a story of men who have lost (or sold) themselves, battling with 'unnatural' men and only defeating them when they adopt their own excessive (and perhaps unmale) forms of masculinity. The pervading sense of sexual infection which crosses the divide between inhabitants and invaders, rural monsters and urban exploiters, rapist and victim, compromises the marked identity of each from the other. It is, then, *anxiety* which characterizes the identity of these men, much more than violence, action or control. The world of the film is almost entirely populated by members of one sex, but paradoxically it is precisely because of this that sexual difference is writ large in every scene. The sodomy scene lays this bare, and the way in which the rest of the film continues to rewrite this event through its consequences provides our way back to the questions of deferred action under analysis here. Ed is tied to a tree by his neck, and must watch as Bobby is stripped. Rather than fighting back, Bobby begs for mercy. 'Hey, boy, you look just like a hog,' says Bobby's assailant, 'Go, Piggy – give me a ride.' The scene then goes one step beyond; Bobby moves from human to pig, from male to female, and exactly what kind of sexual assault takes place at the point of his rape is appallingly ambiguous: 'Looks like we got us a sow here instead of a boar . . . I'll bet you squeal like a pig.' If fantasy disrupts all sense of subjective identification (the subject of 'A Child Is Being Beaten' might occupy any number of active or passive positions in relation to the fantasy, as we have seen), here we are presented with an image of a man astonishingly 'undone' by the identifications of another's desire, which begs the question, 'Is this entirely an act between men?' How much is the degradation of the scene bound up with the feminization of the assaulted bodies? 'He got a real pretty mouth, ain't he?' says the second mountain man, as he moves towards Ed. 'You're gonna do some prayin' for me boy, and you better pray good.' If deliverance is surrender, these are men in part, and on both sides, 'delivered' of one kind of masculinity.

In order to think about how the remainder of the film deals with this event, once it has become 'past', I want to pause on something curious about the way in which the film itself has often been read (and remembered) by critics. A kind of collective forgetting of Bobby's rape prevailed amongst those writing at the time of the film's release. For *Sight and Sound* in the autumn of 1972, 'All the Saturday matinée thrills are there', but nowhere in the review does it mention that one of these is sodomy. Similarly, the *Monthly Film Bulletin* mentions a 'sexual assault' almost in passing and only in the synopsis. The misremembering from which critics have suffered is, however, very like something which is going on inside the film itself. If audiences are desperate to bury and forget, but only find that they need to do this *afterwards*, this is only because those in the film are too, and it may be that one is a strange effect of the other.

The character of Lewis has a lot to say about loss as an active and positive experience. Although, early in the film, he has proclaimed both that he's 'never been lost in [his] life' and that 'there's no risk' ('I don't believe in insurance,' he says to Ed), he nevertheless continues to counsel a certain kind of macho submission: 'sometimes you have to lose yourself before you can find anything.' Yet beneath this lies a more pervasive discussion of loss, partly carried out through the way in which the film deals with its own past, and here too masculinity is at stake. Bobby is raped, and then his rapist is killed, but it is only through the lens of the second event that the implications of the first event begin to come clear. It is, then, only when the men work on the consequences of murder that rape becomes a problem. The second event then causes a gradual rereading of the first, by Bobby, by Ed, and then by Drew. The men respond to Bobby's rape and his rapist's murder through a process of burial, figurative and actual ('This thing's gonna be hanging over us the rest of our lives,' says Lewis). The positive, *active* loss which Lewis counsels (and maybe Bobby gets) turns to a desperate, communal act of repression:[49] as the dead male body becomes a liability, the camera refuses to cut away from it, instead roving around the four desperate men who plot their next move with the body between them. To 'lose' it for good, they dig up the earth with their bare hands, and hide the body in a shallow grave soon to be overtaken by the flood-waters, 'Hundreds of feet deep . . . that's about as buried as you can get.' But at the same time, and only *because* of this, Bobby is enforcing a burial of his rape (in which Ed is also complicit): 'I don't want this getting around'. The act of rape (which figures Bobby as victim) is understood and then buried by *his response* to the act of murder (which figures him as co-aggressor).

A more unnerving (and complex) collective disavowal and revision takes place a little later, following Drew's death, and here the audience too is beginning to be implicated in the deferred understanding. Along with Ed, we see Drew fall, or throw himself, overboard. When the now-injured Lewis suggests that he was shot, immediately our memory of what we saw (did he fall or was he pushed?) is challenged by the men's desperate desire to deny suicide and put murder in its place. The quickness and force of this denial of a more disturbing truth (that Drew's sense of control had terminally 'failed' him) is then symptomatic of their understanding (and hasty revision) of it after the event.

But what is frightening about Drew's death only becomes clear once the second murder of the film is committed. Ed kills the second mountain man (the man he *thinks* killed Drew and nearly raped him), but Bobby still raises the question of Ed's victim's identity further on in the film ('Are you sure that's him? . . . It wasn't just some guy up there hunting?'). Bobby's rape was followed by his rapist's murder, the second trauma which activates in Bobby his anxiety regarding the first. Now Drew's death is followed by the murder of the man who *may* be both Ed's would-be-rapist and Drew's killer (as long as we believe it wasn't suicide), a second trauma which opens up a whole range of questions circulating around how they (and we) rethink the moment of Drew's death. Bobby asks 'Are you sure?', and Ed replies *'You tell me.'* As he speaks an astonishing thing happens. Turning the corpse's face towards Bobby, Ed looks away and in the process turns his own face towards us, looking

49 An act of 'collective psychology' which could benefit from discussion through Freud's later social texts such as 'Mass Psychology and the Analysis of the Ego' – an issue I do not have space to pursue here.

straight to camera and at the audience. The question becomes one for us too, and Ed's direct address, raised by his look, challenges us to make a response which is impossible. *Could* we tell him? Do we *know* what we've seen? Can we be *sure*? We are as lost in these judgements as are those on-screen.

Something curious is therefore going on both outside of the film and within it, concerning what is remembered and what is forgotten – how subjects process the past. Its power lies more in what it tries *not* to show, how it plays with what you saw and didn't see, what it denies you as well as what it offers up. Its truths only emerge to be hastily buried, the second time around, as they are disturbingly revisited by the unfolding of events. This is a kind of misremembering which has infected the way in which the film has been understood by audiences who, like the disavowing subject, have taken away with them something they never actually saw, remembering their own disturbance as if it reflected a visual experience which is rarely there on-screen. Perhaps more significant than this misremembering is the anxious forgetting which has taken place, on- and off-screen. The film itself concludes with Ed's very inconclusive view; his masculinity rests on an ability to repress, so what kind of masculinity we are left with here is indeterminate. One of the final images is of his nightmare of a flooded landscape, a woman's voice saying 'Go to sleep', an arm rising from the waters against the pressure of its burial. These images do not add up to a redeemed, sovereign sexual identity which only sits on *one* side of the great sexual divide, and nor do they testify to a past dead and gone, but one which continues to be reworked.

Somehow, then, it is the men's desperate need to forget the awful things they experience which means, paradoxically, that nothing here is ever entirely finished with. This is an open text, not simply because of its inconclusiveness, its 'failure' to resolve meanings, but in the way it keeps anxiously returning to its own past. Repression and remembering are intimately bound to each other across the different moments the subject inhabits (both the characters on screen and the audience, who are invited into identification with their disruptive sense of 'rewritable' reality). What is interesting is that this process takes place not simply (and crudely) at the level of a character reworking his own anxieties on-screen, but that deferred understanding and the repression which follows it is something which, in the world of this film, takes place collectively, and this is a collective disavowal or rereading of the past which implicates the audience too. Deferred action here is then something which takes place not on the level of unconscious individual response, but as an effect *between* characters, and in a way which also transgresses the boundaries of the film. I am not suggesting that we tackle, say, Bobby's response as a conscious version of the way in which Freud reads the 'Wolf Man' (in the manner of Marie Bonaparte). Instead, we need to be able to account for how these slippages *backwards*, which change the way we read back through a history or a narrative, occur collectively, and infect our ways of reading. For *Deliverance,* then, a kind of deferred action takes place which sweeps the audience along with it, a retrogressive refurbishing which happens in two stages and two places, between moments in the narrative and between screen events and audience response. Peter Brooks discusses the way in which texts involve readers in a form of countertransference: the text's analyst reads the text in terms of his or her own desire, the text 'appeal[s] to [the] complicity' of the reader, who must grasp, 'not only what is said but always what the discourse intends, its implications, how it would work on him'. Whilst I agree that something of this is going on in all reading processes, in *Deliverance* this process is *actively* deployed by the strategies of the film. This is a film which makes you question the terms under which you proceed to make sense of it. In the next section I shall develop these questions of how the audience is implicated, as I piece together the work the reader of Mrs Oliphant's *Autobiography* must do across the fragmented text she presents.

III
Lying Between the Two: Mrs Oliphant and the Narrative Past

What casuists we are on our own behalf! – this is altogether self-defence.[50]

I want to turn now to quite a different text. Mrs Oliphant's encounter with the past which is still present is exemplary, since it sets up a uniquely psychoanalytic dynamic of trauma worked through a rather more complex literary sequence. For both psychoanalysis and, as we shall see, Victorian prose, the past is never lost; it keeps recurring, and it insists on being always rewritten. Mrs Oliphant's *Autobiography* is, however, for the most part an informal collage of reminiscence and the recounting of events, but its tenses, its sense of loss, and its anticipations are severely troubled. This is not just because Oliphant's life itself was so difficult, although it is true that the content of this narrative is at times distressing to read. As the text progresses so Oliphant is systematically bereaved, as her husband, her brother, and all of her children die one by one. Its very mode of telling is troubled by the double difficulty of allowing these traumas to live in the text and then letting them go. If catharsis is a form of positive loss, a letting go which signals release, the 'deliverance' of bereavement is something else – a loss which is not necessarily a

50 *The Autobiography of Mrs Oliphant* (1899) (Chicago and London: Chicago University Press, 1988), p. 7.

freedom. Such events cannot be properly 'past', in the sense that the past is what is excluded from the present. 'All my recollections are like pictures,' she writes, 'not continuous, only a scene detached and conspicuous here and there.'[51] Again, later, she writes of a 'picture' which 'got itself hung up upon the walls of my mind'.[52] There is a sense here that memories have already slipped out of time. In Oliphant's narrative, the past inhabits the present, and is constantly reworked because it is never quite finished; its 'truths' are never fully realized or revealed. Only by working through the past can there be any hope of finding out what it was.

Oliphant writes with a full sense of the traumatic history of a moment thickly layered with the significance of its past and its future which lies already somewhere underneath. Nothing rests in itself: all is anticipation and memory ('too early and too late'). No experience is single, nothing which has sufficient significance happens only once. Things here are always out of time, and this is not just because the double nature of autobiography means that events occur once lived and twice written, recounted, rewritten. Certainly writing here acts like a kind of transference relationship. In the 'Dora' case Freud uses a literary metaphor to discuss transference as the reproduction in analysis of 'new editions or facsimiles' of old impulses and relationships; transference then allows the analysand to produce 'new impressions or reprints . . . *revised editions*'.[53] But even within Oliphant's basic narrative itself there is an incessant referral forwards and backwards. Add to this her repeated awareness of paths *not* taken, parallel courses which might almost be being lived alongside this one, the uncanny journey one *escaped* which can still be identified by tracing back to the moment at the crossroads when another choice was made,[54] and the *Autobiography* presents some astonishing discoveries of the narrative time of the subject – the loops and rereadings which occur when the self re-encounters its own past story. If Phillis carries a past inside her waiting to detonate in the present, Mrs Oliphant, like the 'Wolf Man' and the men of *Deliverance*, can only assimilate the past upon encountering and reworking it a second time.

Steven Marcus writes of Freud as a 'historical virtuoso' in his ability to

51 Ibid., p. 39.

52 Ibid., p. 69.

53 See Freud, *Dora*, pp. 157, 158 respectively; my italics. This is also discussed by Steven Marcus.

54 She writes insistently of the 'might have been', almost as if it is an alternative past also subject to revision: early in the text, looking back, the past is not a place of surety but of even more doubt: 'now I think *if* I had taken the other way . . . it *might have been* better for all of us', and then, 'I *might have* . . . I *should* in all probability . . . I *might* have had . . . they *might* have learned' (ibid., p. 6; my italics). Finally she tries to resolve the matter: 'Who can tell? I did with my labour what I thought best, and there is only a *might have been* on the other side' (ibid., p. 7; her italics).

move between 'the different time strata of Dora's own history . . . back and forth between the complex group of sequential histories and narrative accounts'. In relation to Dora's dreams, this becomes even clearer:

> every dream, [Freud] reminds us, sets up a connection between two "factors," an event during childhood" and an "event of the present day – and it endeavours to reshape the present on the model of the remote past."[55]

Whilst I want to resist analysing Oliphant as one of Bonaparte's textual dreamers, in order to understand the *Autobiography* at all on the level of plot and sequence, we need to be able to move between different moments of psychic history as the dream analyst can. Alongside her sense of retrospection is a constant sense of anticipation, as events in the future leap back into a moment which strictly precedes them. From the start it is difficult to place a narrator who writes in the past tense, of an event which is not to take place until the future, yet who is already able to identify that event *as history*: 'When my poor brother's family fell upon my hands' we are told on page 6, of an event which will only take place in the text on page 123, 'and especially when there was question of Frank's education, I remember that I said to myself . . .'. Although this is an event which has yet to come, it is also already memory. Whilst these are on one level the conventions of first-person retrospective narrative, Oliphant deploys these wrinkles in time to an extreme degree. The result is a narrative remarkable for its temporal uncertainty, as traumatic events erupt before and after their time, refusing to be forgotten or ever properly introduced.

We plunge into these uncertainties, as well as into the midst of bereavement, with the first line of the *Autobiography*:

> Twenty-one years have passed since I wrote what is on the opposite page. I have just been reading it all with tears; sorry, very sorry for that poor soul who has lived through so much since. Twenty-one years is a little lifetime.[56]

We begin in the middle: the story has already started, certain events are quite past, time has passed, even the start of the *Autobiography* itself is done. There is nothing 'on the opposite page' except, in this Chicago University Press edition at least, an informal photograph of Oliphant flanked by the sons and nephew who will not survive the story she is to tell, staring frankly at the camera. The opening is absent, but it also happens twice, once here – the surrogate opening which stands – and then once again where the writing 'on the opposite page' – Oliphant's 'real' opening – has been inserted by the editors of this edition, later in the text, 'so as to preserve the sequence of the narrative' (footnote, p. 3). A choice,

55 *In Dora's Case*, p. 74.
56 *Autobiography*, p. 3.

then, has been made in favour of the century's chronology, for the fragment written twenty-one years earlier, which discusses the death of Oliphant's daughter, actually comes later in the text itself, where it 'ought' to.

I do not wish to discuss the merits of such authorial or editorial decisions,[57] although they are not unlike those which Freud makes in his laying out and retelling of the analysand's story, as a sequence which makes analytic as well as narrative 'sense'. Both of Oliphant's openings are posthumous: we expect autobiographies to start at the beginning, yet neither the opening quoted above nor the fragment to which it refers (which now comes *later*) seem to be what we understand as the beginning. Rather, they start with an awareness of something having already happened, a trauma already in place. The second beginning which ought to be the first – the fragment to which the above lines are referring – begins thus: I did not know when I wrote the last words that I was coming to lay my sweetest hope, my brightest anticipations for the future, with my darling, in her father's grave.'[58] Yet if this is 'really' the start, then it already assumes that something is in place, 'those last words' which are not there yet to which the text is now referring back, as to a moment of ignorance which precedes the awful knowledge which she *now* writes in terms of. Something lies before each beginning: the *Autobiography* is built upon a layering of pain which the text has no choice but to keep returning to and trying to assimilate.

This, then, is a narrative which comes into being with pain in place, and there is no real 'before' to it. It is not a story of a fall from grace, or from joy to sorrow, with a beginning and an end. As an autobiographical subject, Oliphant is born with these textual-future events already in mind (already biographically past), and the style and being of the text is entirely informed and magnetized by them. Just as in psychoanalysis 'it always takes two traumas to make a trauma', so here the trauma happens several times over: as the event, the telling, and the pre-emptive echoes, as well as the aftershocks. Nothing is done with here, and the reverberations of trauma across the time of Oliphant's narrative indicate that subjective events happen *not only* at the time at which they happen.

57 These kinds of issues have, however, begun to attract the attention of psychoanalytic critics. Jane Gallop, for instance, reads the significance of editorial errors and typos as forms of parapraxis – slips of the tongue or the pen (or, indeed, of the keyboard) which can be read for their unconscious resonance. Elizabeth Wright suggests that Gallop 'tries to read not masterfully, but transferentially, attending to slippages of meaning in the text (be they produced by author, typsetter or reader), treating these as effects of "the fading author", resurrected by the desiring reader, for whom the author is the 'subject-presumed-to-know' in the transference relation. . . . the author is uncannily alive, producing ideological effects which require a symptomatic analysis' (*Feminism and Psychoanalysis: A Critical Dictionary*, pp. 226–7).

58 p. 92.

This makes reading difficult. The reader must abandon the certain sense of time passing in only one direction which is offered by other great Victorian biographies (Gaskell's *Life of Charlotte Brontë*, for instance), since always the narrative looks back and forward on itself, and events are weighed with the significance of those past and those to come. On top of this, sequence is jumbled: events never happen only in one place, at one moment in time or in the text. It is often extremely difficult to identify exactly when or where we are, so much does the narrative leap forward to the significance of a future not yet experienced, and backwards to a time which had the present programmed into it as its destiny. It is at times very difficult to gauge the *Autobiography*'s chronology, partly because it was composed in fragments and partly because the anticipatory phrases which pepper the text, pointing its reader forward to events now in the actual past of the writer, but still in the future of the narrative itself ('I did not then know it but. . .', 'as I was to learn', 'as we shall see,' etc.), are subject to an unnerving process of revision. I have said that one of the difficulties of this text is that it was written across a period during which its anticipated audience – Oliphant's children (she writes that she is writing *for them*) – died one by one. What, then, is at stake in our reading of a text which progresses only at the cost of its 'real' listeners being lost, until it finally falls into unfinished silence as Oliphant writes herself to the present?

> And now here I am all alone.
> I cannot write any more.[59]

Where are we placed as the story's audience? Few realist conventions are kept to here: unfinished, diffusely structured, this is an impressionistic and sometimes ragged analysis which rewrites the meanings of subjective history as it rereads individual events. It is also for the most part a text of mourning, as Oliphant recounts and works through her losses in print. The narrative, then, hovers between two moments of trauma: Oliphant's attempted assimilation of the event and the event itself, in between which lies the text and its contradictory reading of that trauma. 'The truth most likely lay between the two', as she tells us of one of these moments, and it is this articulation of a truth 'lying between' the two moments of trauma and analysis to which I now turn.

Here, Mrs Oliphant lays out her discovery of her husband's dishonesty to her, and uncovers her own feeling of separation, loss, anger, as her narrative slips between its different layers of realization. I will quote the passage in full, since it is one of the most important moments in the whole text:

59 p. 150.

> Before we decided definitely to give up everything and go abroad, Frank [Oliphant's husband] went to consult Dr Walsh, who was the great authority on the lungs at that time. He lived in Harley Street, I think. I went with my husband to the door, and leaving him there walked up and down the street till he came out again. I think he was to meet Mr Quain there, who was attending him at the time. And here again there is a moment that stands out clear over all these years. I was very anxious, walking up and down, praying and keeping myself from crying, sick with anxiety, starting at every sound of a door opening. He met me with a smile, telling me the report was excellent. There was very little the matter, chiefly over-work, and that all would be well when he got away. The relief was unspeakable; relief from pain is the highest good on earth, the most exquisite feeling, – I have always said so. It was in the upper part of Harley Street that he came up to me and told me this, and my heart leapt up with this delightful sense of anxiety stilled.
>
> Afterwards, in Rome, Robert Macpherson told me what he said was the true story of the consultation – that the doctors had told Frank his doom; that his case was hopeless, but that he had not the courage to tell me the truth. I was angry and wounded beyond measure, and would not believe that my Frank had deceived me, or told another what he did not tell to me. Neither do I think he would have gone away, to expose me with my little children to so awful a trial in a foreign place, had this been the case. And yet the blessed deliverance of that moment was not real either. The truth most likely lay between the two.[60]

The moment in Harley Street is not a 'primal scene' in the usual sense, although (like that first moment read by Lacan in Poe's 'Purloined Letter') it might be called the *text's* primal scene. This is a story of a deceit, but it is not structured in terms of a simple, two-moment sequence of deceit and revelation, the burial of a mine and its detonation. There are at least three moments here: the deception (if such it is), the revelation by Macpherson, and the retelling and reassessment of the event in the narrating of it. Having read the passage once, it is impossible then to read it a second time innocently. Reading to the end and then looking back, the opening sentence – in fact the whole of the first paragraph – is completely altered in significance. The implication first is that the decision to go abroad depends upon a clean bill of health from Dr Walsh. This changes on a second reading, as it does when Oliphant herself rereads the moment: 'Before we decided definitely to give up everything and go abroad' is a statement only Oliphant, alone, can make: her husband, it seems, has decided to go abroad whether the consultation is good or not. There is the haziness of memory here, and also a certain crystalline clarity, indicating that time distances us from experiences irregularly and with uneven emphasis: 'I think', writes Oliphant once, and then again, but she moves towards certainty: 'And here again there is a moment that stands out clear over all these years.'

60 pp. 46–7.

But the moment changes in the writing of it, as it has already once been changed by Macpherson's revelation. It was once, at its first encounter, a moment characterized by 'unspeakable' relief and 'the most exquisite feeling'. But the moment does not stop there: even if, as a point in time, it has passed, still it continues to be experienced and revisited, powerlessly and painfully. First it is revisited with Macpherson's revelation, and Oliphant, having experienced 'this delightful sense of anxiety stilled', has the relief of the moment stripped from her, so that it is reread in far darker tones. It is not just that her husband lied: it is that the exquisite moment is first given and then taken away, after it has passed. Twice Oliphant underlines the deceit: 'He met me with a smile, *telling* me the report was excellent. . . . It was in the upper part of Harley Street that he came up to me and *told* me this.' By now she is sure of where she was (earlier, she only 'thinks' that he lived in Harley Street), but she is also sure that she is being 'told' a story and expected to believe it. 'Telling' in this context emphasizes the fabrication: this is not itself a real communication from Frank, but a narrative which is open to question. To found one's exquisite moments upon this 'telling' is to risk having those moments dissolve when the telling turns out to be a tale. Frank gives her the relief, but removal of its cause strips her of more than the feeling itself. A key moment is entirely undone: it did happen, but it also didn't, not really. The memory cannot be erased, but it begins to unravel as it is retold: 'And yet the blessed deliverance of that moment was not real either.'

The status of the moment of deceit becomes increasingly questionable as it is retold; there are deferrals here, enacted by a series of narrative layers. Oliphant's own story sets in motion the process of undoing Frank's, yet upon Frank's story part of Oliphant's rests. In telling *her* tale, *Frank's* tale is untold, but undoing Frank's lie means that a gap is created in Oliphant's text. More than in any of these composite elements – the original pain of Frank's illness, the double pain of being lied to, the anger at being left – the trauma of this scene lies in the fact that Oliphant here encounters through narrative an awful moment which slips off the page, which is undone, in the very telling of it. It is past yet still horribly open to question; its remembered pleasure cannot be relied upon, and that pleasure is at the heart of deception. The moment happened and yet it didn't. And as Oliphant grapples her way up the other side of its retelling, she encounters some of the worst elements which it contains – that Frank lacked courage, that he was wilfully deceptive, that he could tell Macpherson but not her, that he could not have known after all because he would never have wilfully exposed her to the difficulties which followed, and finally, that 'the truth most likely lay between the two'.

These are not simply classic stages in the process of mourning, they are moments in the telling and rereading of a tale. This is a narrative which moves, through its loops and revisions, in a pattern like that of the analytic subject in relation to her time and her present-past. In writing a crucial past moment, Oliphant reveals and assimilates that moment through her reassessment and refurbishment of it. She exchanges one form of 'distortion' for another: her husband's protective dishonesty traded in for a later, almost unspeakable distortion of her own. This is not to say that the moment is not one of clarity and realization; rather, that this is a passage written through a hall of mirrors, which proceeds from a point well after the incident itself, as well as her husband's own life, was strictly closed, and then only moves further and further away from its truth even as it carries with it the echoes and reflections of that truth. The 'truth' of this text is beaten out of reflections of an original moment which is long gone but which also remains as memory, a memory which behaves remarkably like a sickness which has lurked in the subject too long, its function and image being now out of all proportion to the germ that it was when it entered the system. What we have is a narrative wincingly negotiating its own bumps and refractions, more telling for its distortions and for what it has become at the end of the hall of mirrors.

But in conclusion I want to go back to my earlier quotation of Oliphant's final line:

> And now I am here all alone.
> I cannot write any more.

Just as the death of her husband gives retrospective significance to the way moments in his life are written by her, so this sense of a text's ending does confer a retrospective reading. In 'Freud's Masterplot' Peter Brooks discusses how, despite the fact that narratives seem to proceed 'linearly, in sequence, in one direction',[61] in life as in text 'the beginning in fact presupposes the end'.[62] Texts, like lives, are infused with what Heidegger called 'Being unto death', but Brooks discusses this through Walter Benjamin and, here, the subject of Jean-Paul Sartre's *Les Mots*, who is,

> determined, as promise and annunciation, by what he would become for posterity. He began to live his life retrospectively, in terms of the death that alone would confer meaning and necessity on existence. As he succinctly puts it, "I became my own obituary." All narration is obituary in that life acquires definable meaning only at, and through, death.[63]

In a very real sense, then, our stories are infused with the retrospective revisionism of a grand as well as a specific *nachträglichkeit*, but – as Brooks argues – the greatest moment of revision, the moment at which

61 p. 282.
62 p. 283.
63 p. 284.

the narrative will be reassessed, understood a second time over, is necessarily always *experienced* in *anticipation*: this is the moment of death.[64] This, then, is destined to be a final moment of revision which only happens outside of one's own narrative, by the 'readers' who survive. In the next chapter I shall shift back to the way in which Freud thinks about how that moment infects the life of the subject through the death drive, and how this is also anticipated in the literature of the uncanny.

64 If, then, the 'death' of a text is read as the prime moment of deferred understanding – the point at which the chain of plot finally makes sense – this must be to some extent true of the subject's story too, except that she would keep the deferral going for as long as possible. The death of the 'author' of one's life hands the possibility of deferred understanding over to someone else.

5

'A Short Way by a Long Wandering': Writing the Death Drive

There is an uncanny moment in Thomas Hardy's 1891 novel *Tess of the d'Urbervilles* when Tess is visited by her own death, not in the shape of a self-spectre returning from the dead future to find the living Tess as its past (this is how Christina Rossetti will encounter death, later in this chapter), but as the simple passing of a sly moment in the calendar year:

> She philosophically noted dates as they came past in the revolution of the year; the disastrous night of her undoing at Trantridge with its dark background of The Chase; also the dates of the baby's birth and death; also her own birthday; and every other day individualized by incidents in which she had taken some share. She suddenly thought one afternoon, when looking in the glass at her fairness, that there was yet another date, of greater importance to her than those; that of her own death, when all these charms would have disappeared; a day which lay sly and unseen among all the other days of the year, giving no sign or sound when she annually passed over it; but not the less surely there. When was it? Why did she not feel the chill of each yearly encounter with such a cold relation? She had Jeremy Taylor's thought that some time in the future those who had known her would say: 'It is the -th, the day that poor Tess Durbeyfield died'; and there would be nothing singular to their minds in the statement. Of that day, doomed to be her terminus in time through all the ages, she did not know the place in month, week, season, or year.[1]

Tess is here marked out as a proto-modern subject, changing 'at a leap', Hardy tells us, 'from simple girl to complex woman' with the realization

[1] *Tess of the D'Urbervilles* (Harmondsworth: Penguin, 1990), pp.149–50.

of her death-date already lurking in her as-yet-unlived history. She is marked by what is called elsewhere 'the ache of modernism', by a sense of her mortality as a historically fixed event of which she will never have a privileged knowledge, already depressed by the clash of scales bound up in an event which will be, and already is, tremendously important for her and equally trivial to everyone else ('there would be nothing singular in their minds'). Elisabeth Bronfen argues in *Over Her Dead Body* that from the start Tess is marked not just as object of exchange in an exogamic kinship structure, but as an object of exchange *'with the dead'*:[2] 'her story becomes an allegory of the process of dying, an interplay of wounding and retribution that occurs at her body'. Bronfen's primary point here is that it is the dead who have chosen Tess, the dead 'who speak through the father's name'[3] and drag her back into a death which is *before* life and bound to the name Durbeyfield. But the moment of Tess's encounter with her image in the glass reveals another aspect of death, or rather a drive which is already in play and regularly punctuating her days. If she doesn't know the day, then it could be any day, every day, and Tess begins to recognize a repetition and a visitation not just of and by her living image in the mirror, but of her dead self already indelibly marked as history. Indeed, 'you're history' is the uncanny message her mirror-image gives: Tess lives on, but in the presence of this newly-claimed 'cold relation' ('not the less surely there', even if she does 'not feel [his annual] chill'), keeping company with the ghosts of D'Urbervilles past.

Tess's sense of a future event returning with the regularity of a birthday, already marked out and already implicating – claiming – her, resonates with a number of positions on death which were to be crystallized in the period at the threshold of which this novel stands. Tess's death is already happening and already being repeated in these silent, *pre-mortem* anniversaries, not just in terms of bodily decay, but in the repetitive messages of her new 'relations'. Death is consequently not just a one-off event, the event to end all events, although this might seem to be what Freud means when he makes the biological point in *Beyond the Pleasure Principle* that 'the aim of all life is death'.[4] Death is a thing which keeps happening, but not just in the form of Tess's metaphysical countdown. The chapter from which this passage comes begins with Roger Ascham's dictum, 'By experience we find out a short way by a long wandering'.[5] This might be a motto for the sense in which Freud's death drive is articulated through 'ever more complicated detours before reaching

2 *Over Her Dead Body: Death, Femininity and the Aesthetic* (Manchester University Press, 1992), p. 233; my italics.

3 Ibid., p. 243.

4 PF vol. xi, p. 311.

5 Quoted by Hardy, *Tess of the D'Urbervilles*, p. 149.

[the] aim of death'[6] – the short way back to the point from which we came, via the long route of life as the deferral of death itself. It is this 'wandering', this psychic process, which plays out a mutual desire for, and a fending off of, death, which is the subject of this chapter. At the heart of structures of repetition, the death drive is 'found out' by betraying itself *in* repetition, known through the flight of the drive toward death, through its self-betrayal in life, through its work on the living subject and its text.

The texts on which this chapter focuses foreground this drive in its various forms: the mechanisms of repetition encountered here by Tess (but also articulated more formally in a plethora of other texts we could choose); the projection into liminal 'undeath' explored in vampire literature; the ascetic or masochistic self-undoing of Christina Rossetti's lyrics which posit the subject in a moment carefully etched out beyond the pleasure principle. Such texts are not, however, concerned with actual death, but with its drive, the drive toward an object which can never be 'cashed in' by the subject in the form of arrival at and satisfaction in death as the drive's object. Once the object is reached, the experience of the subject inevitably ceases. Thus these various death narratives explore the possibility that long before death is reached, it sends out pre-emptive echoes of itself (like Tess's silent death-day anniversaries), a form of inverted *déjà vu* characteristically formed as the repetitions of psychic life.

I

'The Aim of All Life': Freud's Death Drive

Freud's theory, that human life builds towards, and is motivated by, its final desire for death, was developed during and after the First World War as the result of certain puzzling phenomena which he had by then encountered: the compulsive repetition of painful events (particularly in the dreams of victims of war neurosis) with no therapeutic resolution; the repetition and recreation in some analytic situations of the patient's most disturbing experiences, accompanied by an active blocking of the process of working through those experiences, and a resistance to cure (the negative therapeutic reaction); the problem of masochism, within which the subject takes pleasure in pain culminating in the final loss of self; and the child's 'fort-da' game which, through staging a game of loss and recovery of a cotton-reel, emphasized for the child the moment of loss over the moment of recovery of the mother. Each of these underlined the sense that there is pleasure to be had in trauma and that, rather than

6 *Beyond the Pleasure Principle*, p. 311.

wanting egoistic fulfilment, the subject is finally bound to want a dissolution, breakdown or abnegation of the egoistic self. Desire is, then, not necessarily the agent of life instincts; it also, or perhaps fundamentally, serves the death drive – desire culminates in the extinction of the subject ('The pleasure principle seems actually to serve the death instincts,' as Freud puts it in the final paragraph of *Beyond the Pleasure Principle*).

In 1914, long before the theory of the death drive was developed in any coherent form, Freud's work began to change drastically with the publication of his paper on narcissism, which offered a sexual model which turned erotic impulses back onto the self in a circular form of desire which needed no external object. Narcissism is important in this genealogy of the death drive in that it shows sexual drives being directed toward the self, with the ego as object, which revealed something of the drive's general will to return. Certainly, a psychic darkness is already evident in the painfulness of each of the behavioural possibilities outlined above; compulsive repetition, masochism, the fort-da game, all stress and enforce disturbance or trauma rather than the straightforward pursuit of pleasure. In tandem with this, Freud identified a pattern of psychic circularity at work, in these painful repetitions, in the masochistic focus on self-traumatization, in the going and returning of the cotton-reel, in narcissism as a model for desire deflected back onto the self. It is through these various forms of return that the death drive, which urges the subject toward an 'original' moment rediscovered in death, becomes visible.

If narcissism is the first circle (within which desire orbits back to the self it came from, in self-love), then a second circle is built upon the findings of the narcissism paper by the unresolved questions Freud posed during the war concerning the nature of psychic repetition. The first reference to repetition compulsion (*Wiederholungzwang*) in Freud's work comes in the 1914 paper on technique 'Remembering, Repeating, Working-Through', where it is understood as part of the process of healing; compulsion to repeat is the patient's 'way of remembering'.[7] The repetition in analysis which can take place through the transference relationship is *aided* by a phenomenon which Freud is eventually to think of as a component and symptom of the death drive. In transference 'the patient yields to the compulsion to repeat':

> We soon perceive that the transference is itself only a piece of repetition, and that the repetition is a transference of the forgotten past not only onto the doctor but also onto all the other aspects of the current situation.[8]

[7] *SE*, vol. xii, p. 150.

[8] Ibid., p. 151. Peter Brooks writes: 'Repetition is both an obstacle to analysis – since the analysand must eventually be led to renunciation of the attempt to reproduce the past – and the principal dynamic of the cure, since only by way of its symbolic enactment in the present can the history of past desire, its objects and scenarios of fulfilment, be made known, become manifest in the present discourse' ('The idea of psychoanalytic literary criticism', p. 10).

However, by 1920 Freud is writing that it is 'the compulsion to repeat which first put us on the track of the death instincts'.[9] Indeed, what repeats for Freud during the war is horror, trauma, shock; not only was he confronted with patients who compulsively and masochistically returned to experiences which only caused them further, unresolvable pain (for instance, the repetition of war traumas, which did not apparently build toward any kind of therapeutic working through), but repetition *per se* was itself a questionable experience. Thus the death drive is present not simply in its content (the horror of whatever repeats) but in the fact of repetition itself. Again, in *Beyond the Pleasure Principle* Freud notes that children 'never tire of asking an adult to repeat a game' or story. The child

> will remorselessly stipulate that the repetition shall be an identical one and will correct any alterations of which the narrator may be guilty . . . repetition, the re-experiencing of something identical, is clearly in itself a source of pleasure.[10]

However, Freud also notes that 'Novelty is always the condition of enjoyment', and that the compulsion to repeat 'give[s] the appearance of some 'daemonic' force at work':

> when people unfamiliar with analysis feel an obscure fear – a dread of rousing something that, so they feel, is better left sleeping – what they are afraid of at bottom is the emergence of this compulsion with its hint of possession by some 'daemonic' power.[11]

These individual repetitions do, however, build up into a far larger image of psychic life locked into a process of return, the core of the death drive. A few key quotations from *Beyond the Pleasure Principle* will serve to illustrate this famous point. First, Freud employs Barbara Low's term 'Nirvana principle' to describe that 'dominating tendency in mental life . . . to reduce, to keep constant or to remove internal tension due to stimuli'.[12] Toward the end of the essay Freud writes more boldly that 'the most universal endeavour of all living substance [is] to return to the quiescence of the inorganic world'. Repetition and return are thus pro-

9 Freud, *Beyond the Pleasure Principle*, p. 329.
10 This and subsequent quotations, ibid., pp. 307–8.
11 Actually, the demons must already be in some sense awake for this fear to exist. In 'Analysis Terminable and Interminable' Freud writes: 'The warning that we should let sleeping dogs lie, which we have so often heard in connection with our efforts to explore the psychical underworld, is peculiarly inapposite when applied to the conditions of mental life. For if the instincts are causing disturbances, it is a proof that the dogs are not sleeping; and if they seem really to be sleeping, it is not in our power to awaken them.' (*SE* vol. xxiii, p. 231).
12 p. 329.

grammed into organic substance, and the death drive charts the psychic consequences of this. Pleasure, then, gives off into discharge of excitation, the moment of the little death at which the subject is extinguished in orgasm, that concept so beloved of Georges Bataille, the idea of orgasm as 'the little death' as evidence of the death drive:

> We have all experienced how the greatest pleasure attainable by us, that of the sexual act, is associated with a momentary extinction of a highly intensified excitation. The binding of an instinctual impulse would be a preliminary function designed to prepare the excitation for its final elimination in the pleasure of discharge.[13]

Instincts thus originate in *'a need to restore to an earlier state of things'*,[14] and consequently 'the aim of all life is death',[15] or, as he puts it in 'Thoughts for the Times on War and Death', 'everyone owes nature a death and must expect to pay the debt'.[16] It is this sense of a final desire to 'return' which interests me in its connection with Freud's economic model of final discharge at an ecstatic point of zero excitation, for not only did the coming together of these ideas – of return, and of annihilation – require Freud to posit a more developed theory of masochism, they also locked into a wider cultural concern with repetition and return emerging from modernist aesthetics and Nietzschean philosophy.

Freud's philosophical debts and interests are great. As well as aligning his work with Schopenhauer, Kant and Empedocles in various places, here his theory of the death drive as one of psychic return echoes perhaps the earliest Western philosophical statement, Anaximander's fragment:

> And the things from which existing things come into being are also the things into which they are destroyed, in accordance with what must be. For they give justice and reparition to one another for their injustice in accordance with the arrangement of time.[17]

Many of Freud's statements on the economic justice of the death drive directly echo this, but in addition it is also impossible to read the death drive without reference to Nietzsche's theory of the eternal recurrence. In 1917 Freud, under the superstitious impression that his life was to end in February 1918, wrote in response to fellow-analyst Sándor Ferenczi's protests about this notion,

13 pp. 336–7.
14 p. 331; his italics.
15 p. 311.
16 PF vol. xii, p. 77.
17 Quoted by Jonathan Barnes in *Early Greek Philosophy* (Harmondsworth: Penguin, 1987), p. 75.

> When I read your letter I looked down on your optimism with a smile. You seem to believe in an "eternal recurrence of the same" and to want to overlook the unmistakable direction of fate. There is really nothing strange in a man of my years noticing the unavoidable gradual decay of his person.[18]

The reference to Nietzsche is doubly significant here. Freud is engaging in an active misreading of a model which was to be reappropriated a little later in *Beyond the Pleasure Principle*; here, eternal recurrence is taken to mean the unchanging, interminable repetition of the daily events of one's life, of the fact that one *is* alive, against a 'fate' which only moves in one (linear) direction. The 'gradual decay of the person' is then at odds at this stage with 'the eternal recurrence of the same': one moves forward *into* decay, *away from* recurrence, away from 'the same' of health and life. Ageing breaks the pattern of this process of returning – change here is not an agent of repetition, but of irrevocable breakdown.

Three years later Freud was still very much alive and *Beyond the Pleasure Principle* was published, which, as he put it, 'steered [psychoanalysis'] course into the harbour of Schopenhauer's philosophy'.[19] Posing a new 'dualistic' model of the drives, Freud writes:

> May we venture to recognize in these two directions taken by the vital processes the activity of our two instinctual impulses, the life instincts and the death instincts? . . . For [Schopenhauer] death is the 'true result and to that extent the purpose of life', while the sexual instinct is the embodiment of the will to live.[20]

However, this steering of psychoanalysis 'back' into philosophy also has the effect of making Freud an interpreter of philosophy. Whilst a Nietzschean reading of the death drive is important to an understanding of Freud's final economics, at the same time Freud was an important reader of Nietzsche's doctrine of the eternal return, even if this reading is to be found between the lines of Freud's post-1920 work, in half-quotes suggesting a model of desire and return which only rarely overtly speaks its debt to Nietzsche. By the time we get to this phase in Freud's writing, we need to some extent to read Freud as a reader of Nietzsche, and Nietzsche *in* Freud. Between the two poles of, on the one hand, the purely clinical evidence of compulsive repetition and, on the other, the abstraction of the philosophical doctrine of the eternal recurrence, lies the death drive as a model of desire, a desire which wants its discharge, which returns to itself in order to extinguish itself. Here, then, the clinical evidence with which Freud was faced (which could not be explained

18 Quoted by Ernest Jones, *Life and Works of Sigmund Freud* (Harmondsworth: Penguin, 1984), p. 441.
19 *Beyond the Pleasure Principle*, p. 322.
20 Ibid.

solely in terms of life instincts) came together with the model of a returning desire which Nietzschean philosophy was able to offer. This moment of coalescence is the death drive, or at least the death drive in one of its clearest forms, that of the primarily economic and so-called 'Nirvana Principle'.

The death drive was consequently theorized partly because dangerous pleasures taken alone were psychic phenomena which would never finally 'fit' with Freud's early model of instinctual life based on pleasure. In tandem with this was an increasing recognition that human drives are capable of finding internal as well as external objects upon which libidinally to fix, turning back to the self as well as out to the other. In the 1905 text *Three Essays on the Theory of Sexuality* it was the sexual drive which, as Laplanche puts it in *Life and Death in Psychoanalysis*, was 'that drive par excellence' and 'which represents the model of every drive'.[21] However, the object of the drive becomes increasingly problematic, increasingly obscure, the nearer that object of desire is to the desiring self: with Freud's work during the period of the war it becomes clear that drives are not simply aimed outwards toward external objects of desire, but more fundamentally projected inwards, in narcissism and masochism. Finally, with the theory that the death drive is the final 'drive par excellence', it is the death of the self which is the fundamental object of the drive.

The death drive consequently surfaces from within psychoanalysis after 25 years which prioritize sexuality as the basis of all drives. '[S]uddenly in 1920', writes Laplanche, death

> emerges at the centre of the system, as one of the two fundamental forces – and perhaps even as the only promordial force – in the heart of the psyche, of living beings, and of matter itself. [Death becomes t]he soul of conflict, an elemental force of strife, which from then on is in the forefront of Freud's most theoretical formulations.[22]

With the turning of Freud's thought up to and during the First World War, sexuality begins to give way to something different. Having 'done such yeoman service', as Jones puts it, the pleasure principle 'was now stated to be the handmaid of the death instinct'.[23] One final image presents the pleasure principle as only the substance which, mixed with death, allows it to be perceived and read: the death drive 'escapes detection unless its presence is betrayed by its being alloyed with Eros.'[24]

These are concepts thick with literary significance and potential, yet the death drive has itself received comparatively little attention from

21 p. 8; his italics.
22 pp. 5–6.
23 Jones, *Life and Works of Sigmund Freud*, p. 508.
24 Freud, *Civilization and Its Discontents* (1930), PF vol. xii, p. 313.

critics and readers, growing interest in the relationship between literature and psychoanalysis notwithstanding. This is partly because, as I shall discuss toward the end of this chapter, the theory of the death drive, though crucial to late Freud, has been neglected by the psychoanalytic establishment itself, and cultural critics have tended to take their lead from the early Freud. The remainder of this chapter will largely consist of readings which serve to open up the uses of the death drive in literary analysis, involving discussion of fictions which demand a theory of the death drive, even if Freud himself acknowledges that the theory itself might only be a fiction, a 'speculation' (as he does in *Beyond the Pleasure Principle*). I shall focus first on that exemplary image of death 'alloyed with Eros': the vampire.

II

'Girls are caterpillars while they live in this world': *Carmilla*, Insects and Instincts

Oh Rose, thou art sick!
The invisible worm
That flies in the night,
In the howling storm,
Has found out thy bed
Of crimson joy:
And his dark secret love
Does thy life destroy.

William Blake, 'The Sick Rose'

J. Sheridan Le Fanu's 1872 lesbian vampire novel *Carmilla* is an extraordinary text for discussions of the death drive. At one point the seductive eponymous vampire – one of the most articulate in the sub-genre – asks the young woman she is in the process of seducing,

> 'You are afraid to die?'
> 'Yes, everyone is.'
> 'But to die as lovers may – to die together, so that they may live together. Girls are caterpillars while they live in this world, to be finally butterflies when the summer comes; but in the meantime there are grubs and larvae, don't you see – each with their peculiar propensities, necessities and structure.'[25]

What is most interesting about this is not its plea for a romantic fulfilment in death, but its insistent erotic entomology – a love that never dies

25 J. Sheridan Le Fanu, 'Carmilla', in *The Penguin Book of Vampires*, ed. Alan Ryan (London: Bloomsbury, 1991), pp. 95–6.

expressed as the culmination of insect life, or insect development used as an image of erotic change. Carmilla's suggestion is both that it is death which is 'summer' – rather than the termination of all growth, it is its culmination – and that death (or rather 'undeath') must be perceived as another, more fulfilling phase of existence.

Carmilla has a disturbing ability to look at death from quite another point of view, so that it becomes a state with its own positive qualities and attributes: 'life and death are mysterious states,' she says, 'and we know little of the resources of either.'[26] Rather than being the termination and negation of all the good things in life, death is the fulfilment of life's pleasures, but in a radically un-Christian way – death is a woman's summer. Indeed, even before Carmilla's vampirism is revealed, she says a lot of things which indicate her vampiric commitment to the realm of the undead, a liminal space within which the subject shifts into another gear, and ceases to define itself according to the either/or choices of the binary, waking world. Many of these statements can be mapped directly onto the death drive. Here, Carmilla's image of a better life for women does something other than simply assert the possibility of vampiric immortality (an alternative life after death also to be gained through communion of the blood). Carmilla suggests that women's fulfilment is to be had through a developmental route which is anti-human and projected toward the liminal.

The stages of desire, and the stages of the life of women, are then expressed through an image of the transformations of insect life. Blake's 'invisible worm/ That flies in the night' could be read as an early figure for the vampire, bringing a sexualized wasting and death through a disease which infects the victim from the inside out. This worm is a night creature, a shape-shifting, dissolving entity – it contaminates the blood with a 'dark secret love' which ravages the healthy self and activates a depraved range of appetites. We are in the realm of perverse love, foreshadowing the sadomasochistic image of the vampiric invader as both 'worm' and flying thing (caterpillar and butterfly) which Carmilla describes – and, indeed, which Carmilla is.

But the burden of her image also offers a kind of feminist version of the death drive. If for Freud 'the aim of all life is death', then Carmilla will feminize this; more specifically, the aim of all *feminine* life is 'undeath', and it is only in undeath that femininity finds its fulfilment. Having much in common with feminisms focused by an idealized celebration of the pre-Oedipal (discussed above in Chapter 3), Carmilla's feminism – if such it is – is resolutely post-mortem. Beyond this shocking preference – for death over life, or rather for 'undeath' over both – there is also a point to be made about development. Women do not live the seven ages of man; rather, they change like insects do, and their trajectory ignores the

26 p. 96.

division of life and death – indeed, it projects entirely beyond death, so that 'summer' and the true flight of fulfilment takes place somewhere entirely other. It is not because girls are young that they 'are caterpillars' – as yet underdeveloped, unformed and unable to 'take off': Real life is given to be something which happens to females after they have quit 'this world' altogether. Ideal femininity pushes toward a winged state – the 'summer' of feminine development – which transgresses a fundamental boundary.

Vampires are shape-shifters, but, as Carmilla suggests, they do not only fly as bats. Ernest Jones writes in his early psychoanalytic study, *On the Nightmare*,

> Of especial interest is the widespread belief that the Vampire can appear in the form of a snake, of a butterfly, or an owl, for these were originally figurative symbols of departed souls, particularly the parents. . . . When, for instance, one has anything to do with the body of a Vampire, one has to take special care to watch whether a butterfly flies away from it; if so, it is important to catch and burn it.[27]

Whilst *Carmilla* is probably drawing on this traditional vampiric incarnation, the point in the novel is a feminized one, with the stages of insect development standing for the ideal phases of a woman's life. Burn the butterfly and you destroy a woman who is finding her 'summer'. But a further point about development needs to be raised here. For body-horror film-maker David Cronenberg (director of *The Fly* and the insect-laden *Naked Lunch*), insect life is fascinating because it is so entirely other, yet it can offer an image for change which more familiar life forms cannot. This is Cronenberg on insects:

> They don't start small and grow bigger . . . [they] really transform from one kind of life to another – the caterpillar to butterfly, I suppose, is the common example. . . . When an insect transforms, is it really the same insect or is it really . . . something else? Is there some core of identity that still remains, from the egg to the caterpillar, to the pupa, the butterfly itself? Is it really the same, or has something died and something else been created?[28]

Perhaps the literary insect-transformation which exemplifies this is Kafka's *Metamorphosis*, which also asks if the bug is the same as the man and, moreover, what is the power of his sudden change (the central figure wakes up one morning to find that he is a giant bug, and the story traces his gradual breakdown in his unhappy relationship with a world which is even more disturbed than he is). Unlike the eponymous hero of

27 (London: Hogarth Press, 1931), p. 107.
28 Damon Wise, 'Doctor Benway Rides Again: David Cronenberg Interviewed', in *Shock Xpress*, ed. Stefan Jaworzyn (London: Titan, 1991), p. 66.

Cronenberg's *The Fly*, who is – as the character Seth Bruncle (Jeff Goldblum) himself puts it – 'striken by a disease with a purpose' which is the agent of a gradual undoing by the insect-enemy within, Kafka's insect-hero does not *gradually* become less of himself and more of the bug. Although vampires are more often characterized as bats or cats or dogs than the insect of Carmilla's extraordinary image, they nevertheless make us ask a similar question of sameness and difference, and of metamorphosis. Vampirism is also a disease with a purpose, and its purpose is reproductive. The change at stake (as it were) here is underpinned by a further anxiety, that of a reproductive power which bypasses sex, a possibility of transformation and multiplication which does not appear to depend on normal forms of intercourse.[29] Cronenberg's character 'Brundle-fly' is similarly created by the abnormal genetic splicing of two separate creatures (Seth Brundle and a housefly who, unnaturally mixed together, make a third thing). As Brundle-fly himself puts it, 'we weren't even properly introduced'.

Vampires also 'don't start small and grow bigger . . . [they] really transform from one kind of life to another', and in a way which ensures maximum shock-value: the person you thought you knew is no longer that person. Lucy in *Dracula* becomes 'a nightmare of Lucy . . . the whole carnal and unspiritual appearance, seeming like a devilish mockery of Lucy's sweet purity'. 'Is that really Lucy's body, or only a demon in her shape?' asks her lover,[30] to which Van Helsing replies, 'It is her body, and yet not it'. By the next page of *Dracula*, Lucy has become 'the Thing' and an 'it', no longer identifiably a woman (or perhaps too much of a woman).

Carmilla pushes the issue even further. She is a 'summer' woman, no longer contained by the boundaries of girlhood; she can fly, in several senses of the term, moving in ways that are not humanly possible, and moving herself and others sexually. The vampire's flight, like the witch's, can of course be read in quite crude symbolic terms; as Freud himself put it in an early letter to Fliess, 'Their "flying" is explained; the broomstick they ride is probably the great Lord Penis.' Whilst Carmilla, the prototype lesbian vampire, shows little interest in his lordship, the connection between flight and *jouissance* is familiar.[31] But her ease of movement, also suggests that Carmilla has evolved into something entirely other than 'human' (and without recourse to 'normal' intercourse). Like all good vampires, she is anything but stable, shape-shifting with an uncanny changeability, from solid into vapour, from languid passivity to feline

29 I am grateful to Nick Davis for making the point that these notions might originate moment prior to a biological understanding of insect reproduction.

30 Bram Stoker, *Dracula* (1897) (Harmondsworth: Penguin, 1979), p. 256.

31 As was discussed in relation to Carter's Fevvers in Chapter 3, Freud also discusses the amorphous sexuality of dreams of flying in *The Interpretation of Dreams*, pp. 374–7 and 516–17.

predation. She is 'a dream of something black coming round my bed';[32] often 'she' is a fevered haze, a cloud of gas or an exhalation of breath, varying her forms with a disturbing freedom:

> I saw a large black object, very ill-defined, crawl, as it seemed to me, over the foot of the bed, and swiftly spread itself up the poor girl's throat, where it swelled, in a moment, into a great, palpitating mass. . . . The black creature suddenly contracted toward the foot of the bed, glided over it, and, standing on the floor . . . I saw Millarca.[33]

The thing is the same thing but also a quite different thing, and whilst its stable identity is problematized by the thing's sheer versatility, it nevertheless has a grammatical consistency throughout this account: it is a changeable 'it' which is then accounted for by the apparition: 'I saw Millarca'.

Demons have had a starring role in psychoanalytic thought. The vampire can stand as a figure for that demonic unconscious which we encountered earlier in Freud's image of compulsive repetition, characterized by its 'hint of possession by some "daemonic" power' (that 'obscure fear – a dread of rousing something that . . . is better left sleeping'). Coming from darkness, the vampire is more at home in the material world that the Christian soul, implicating her victim in a circle of vampiric hunger which breeds the reproduction of that hunger. She belongs here, and she can move around in a number of ways. It is easy enough to read the vampire crudely as an allegory for a repressed unconscious life which has finally exploded in a particularly dangerous form – the readings of *Dracula* which take this line are numerous. A less vulgar Freudian reading would allow the text rather more potent ambiguity. To shift across to Shoshana Felman's celebrated account of 'vulgar' psychoanalytic readings of James's *The Turn of the Screw* and her own (less vulgar) evocative Lacanian account of the interpretative possibilities of 'the very textuality of the text',[34] what we could call for in the spirit of Felman is a reading which 'would thus attempt not so much to capture the mystery's solution, but to follow, rather, the significant path of its flight'[35] – a reading, then, which flies (like Carmilla) into unbounded uncertainties. It is not only insects which fly in the afterlife; texts, in their critical reception, can too. Reading is then a space of reanimation (or even of deferred action): the text lives one life (as it is written), and then it lives another (when it is read). Reading can do one of (at least) two things with a text: pinning down the butterfly to a range of fixed symbols (the 'nailing down' which D. H. Lawrence argues so vehemently against in his essays

32 'Carmilla', p. 104.
33 Ibid., p. 130.
34 Felman, 'Turning the Screw of Interpretation', p. 117.
35 Ibid., p. 119.

on the novel and in *Apocalypse*: 'once a book['s] . . . meaning is fixed or established, it is dead'[36]), or letting it show what it can do, playing on the reader in a relationship of transference. We can read in a way which allows the text to 'take flight'[37] or we can pin it down to a particular bodily form or single meaning, so that it is reanimated as only one thing. The vampire who 'flies in the night' is then a rather tortuous image for a text which finds a multiple life in the afterlife of reading: turning into a range of life-forms which were not possible in its first incarnation at the point of writing. After the death of the author might follow the undead text . . .

III
Rossetti, Masochism and the Undead Lyric

Christina Rossetti also wrote about flying insects, and as an image for female fulfilment beyond death, but in Rossetti's verse Freud's Great Lord Penis – the thing that makes you fly – turns into the Lord of Christianity. The speaker of the poem 'Two Thoughts of Death' sits on the grave of the woman whose life 'was to have been one with mine' and laments her corruption:

> Foul worms fill up her mouth so sweet and red;
> Foul worms are underneath her graceful head;
> Yet these, being born of her from nothingness,
> These worms are certainly flesh of her flesh.[38]

If here the worms have taken in and have become the woman's flesh, by the second stanza they have turned into a more ethereal incarnation of the woman. Feeding from her body, the worms metamorphose and take flight:

> And lo my hand lighted upon heartsease
> Not fully blown: while with new life from these
> Fluttered a starry moth that rapidly
> Rose toward the sun: sunlight flashed on me
> Its wings that seemed to throb like heart-pulses.
> Far far away it flew, far out of sight, -
> From earth and flowers of earth it passed away
> As though it flew straight up into the light.

36 *Apocalypse* (1931) (Harmondsworth: Penguin, 1979), p .4.
37 To use Felman's term – see 'Turning the Screw . . .', p. 119.
38 *Selected Poems* (Manchester: Carcanet, 1984), pp. 41–2.

The moth flies from the corpse taking the dead woman's spirit with it, and Rossetti, like Carmilla, reads the flight as the image of the woman's fulfilled immortality:

> Then my heart answered me: Thou fool to say
> That she is dead whose night is turned to day,
> And no more shall her day turn back to night.

We are, of course, in rather different theological universes here. Carmilla urges her girlfriend to join her in vampiric death as an act of flight and metamorphosis into a more powerful *bodily* phase, whilst Rossetti's speaker celebrates the dead woman's escape from this, although accompanied by a similar interchange of day and night. The flight of the insect does for both: a flight *into* the body or *away from* it – sexual fulfilment or spiritual transcendence. Rossetti's moth is the last image of corporeal life, hurling itself toward the burning flame in an act of more respectable Christian transcendence. If the vampire-butterflies of Jones's account need catching and burning because they too are infected, Rossetti's incorporated moth-souls will do it themselves, suicidally flying into their own heavenly pyre. Fire is often the agent of transcendence in Rossetti, as with the liminal soul of 'A Pause' who, waiting for the metamorphosis, finds it when 'the tardy sand/ Of time ran golden; and I felt my hair/ Put on a glory, and my soul expand'.[39]

Yet there is a streak of vampirism in Rossetti's work which I want to explore in terms of some further thoughts on the literary death drive. This might seem a far-fetched claim for the work of one of the nineteenth century's greatest Christian poets, but Rossetti was also neice of John Polidori, author of *The Vampyre* (1819), the first fully formed vampire tale published in English. Despite the fact that study of her works is fast becoming one of the biggest growth industries in feminist criticism, it is only in the last year or so that this vampirism has been traced.[40] 'Goblin Market' in particular has been read as a text which deals with this excessive orality and appetite, even if it is not a full-blown account of haemo-

39 Rossetti, 'A Pause', ibid., p. 47.

40 See in particular David F. Morrill's '"Twilight is not good for maidens": Uncle Polidori and the Psychodynamics of Vampirism in "Goblin Market"', *Victorian Poetry*, 28/1, pp. 1–16, which reads this excessive poem of orality in relation to the work of Polidori. Nina Auerbach gives the 'lushly helpless' life-in-death woman a new twist when she reads the Freudian hysterics of *Studies on Hysteria* together with their 1890s vampiric sisters in Stoker, in 'Magi and Maidens: The Romance of the Victorian Freud', in *Writing and Sexual Difference*, ed. Elizabeth Abel (Brighton: Harvester, 1982), pp. 111–30.

sexuality.[41] Lizzie and Laura's oral sexual connection in 'Goblin Market' could be mapped fairly easily onto that of Carmilla and (the other) Laura, even though Rossetti's vampiric sex scene can also be read as the taking of the sacrament:

> 'Did you miss me?
> Come and kiss me.
> Never mind my bruises,
> Hug me, kiss me, suck my juices
> Squeezed from goblin fruits for you
> Goblin pulp and goblin dew.
> Eat me, drink me, love me;
> Laura, make much of me.'[42]

The sadomasochistic orality of this would be easy to decode in simple Freudian or Kleinian terms, but what I want to discuss instead is Rossetti's encounter with vampirism as the transgression of boundaries. In *Carmilla* boundaries between states are seldom certain: 'dreams', like vampiric identities, 'come through stone walls, light up dark rooms, or darken light ones, and their persons make exits and their entrances as they please, and laugh at locksmiths'.[43] She even appropriates other people's dreams to cover her tracks, countering their memory of her night-visits by claiming the vampiric moment as primarily a function of *her* dream.[44] Rossetti's lyrics ignore or play across the life/death divide in a similar way, her voices also making exits and entrances as they please, sketching out a relationship of identification with the dead which Ernest Jones argues is a key component of vampire anxiety. One of the central concerns here is the ambivalence of 'undeath', read as a cursed beyond-life-sentence or as the gift of corporeal eternal life: as the advertising legend of Francis Ford Coppola's 1992 film *Bram Stoker's Dracula* asserted, 'Love Never Dies'. Rossetti's 'A Pause' is only one lyric which takes the point of view of the corpse, so much so that, to appropriate a little from Ernest Jones on this, 'it sometimes needs an effort to remember that [Rossetti] can only represent ideas projected onto [the corpse] from the minds of the living'.[45] One of the slipperiest aspects of her sparse, vertigi-

41 A term used by Christopher Frayling to signify the desire and erotics of blood: 'the most apt general term to describe the sexual basis of the vampire relationship'. He continues: 'Whether vampirism is related to civilization and its discontents (Freud), to suppressed memories in the collective unconscious (Jung), to breast-feeding and the projection onto others of the need to bite (Melanie Klein), or to monstrous manifestations of eroticism for any other reason, I have chosen "haemosexuality" as the most apt general term to describe the sexual basis of the vampire relationship' (*Vampyres: Lord Byron to Count Dracula*, ed. by Christopher Frayling (London and Boston: Faber, 1991), p. 388)

42 *Selected Poems*, pp. 80-97

43 'Carmilla', p. 102.

44 See ibid., pp. 85–6.

45 *On the Nightmare*, p. 99.

nous lyrics is this unfixable quality: a clear voice coming from the marvellous, impossible space.

Something fundamentally 'undead' is then evoked by Rossetti's lyric liminality, even if this can also be interpreted and reappropriated as the Christian doctrine of 'soul sleep'. For Jerome J. McGann, soul sleep 'pervades the work of her greatest years as a poet' and focuses upon, according to the *OED*, 'the state in which (according to some) the soul sleeps between death and the day of judgement'.[46] A range of other short lyric poems address this state, but rather than following McGann's route, I am more interested in tracing their unorthadox positioning. In 'After Death', the speaker, both lying under the shroud and somehow also hovering above the scene, addresses an unhearing male presence, and manages to speak from the point of view of life *and* death. He thinks she 'slept/ And could not hear him, but I heard him' – she has access to both sides of the life/death divide, lying within the scene and hovering above it, seeing it although her eyes are covered.[47] This is a poem of radical miscommunication, mismeeting: they occupy different universes, but hers offers her rather more power than life had done: she is dead – presumed impotent – and yet it is only now, as this ability to witness two different worlds suggests, that she can have any power. Bound up in the pathology of this self-abnegation is also an image of the woman as Christ-the-victim, an identification which is dripping with self-glorification as well as masochism. In 'Remember', the living speaker projects herself so far

46 Quoted in Jerome J. McGann's interesting analysis of Rossetti's poem 'Up-Hill', which is itself read in relation to the doctrine of Soul Sleep. McGann's emphasis is on Soul Sleep as an enabling idea ('no other idea generated such a network of poetic possibilities for her verse' (p. 243)), a moment from which the soul can escape knowingly from the world's corruption and experience, in a pre-emptive manner, a glimpse of the next phase: 'Several of Rossetti's poems set forth paradisal visions, and in each case these proceed from a condition in which the soul, laid asleep, as it were, in the body, is permitted to glimpse the millennial world. . . . The closest approximation one can arrive at in this world to the vision that can be expected after death in Soul Sleep is a description not of paradise itself, but of the emotional effect which results from the actual desire for such a vision' (pp. 244–5). McGann's emphasis is thus on the soul looking *forward* at this moment, whilst my reading explores its obsessive concern with what it has left, or its momentary, suspended inability to move in one direction or the other (as with the speaker of 'A Pause', whose 'soul, love-bound, loitered on its way'). For Angela Leighton, however, 'it is not heaven but entombment which fascinates Rossetti; it is not "Soul Sleep" which characterizes the state of death for her, but a disturbing sleeplessness of the mind and accompanying corruption of the body . . . a prolonged and indefinite twilight of lingering consciousness and physical decay'. It is this compulsively repeated 'inability to experience death' which, in the aforementioned reading of 'Goblin Market', David Morrill identifies as vampiric. See Angela Leighton, '"When I am dead, my dearest": The Secret of Christina Rossetti', *Modern Philology*, 87, pp. 373–88; and Jerome J. McGann, 'The Religious Poetry of Christina Rossetti', in *The Beauty of Inflections: Literary Investigations in Historical Method and Theory* (Oxford: Clarendon Press, 1985). I am grateful to Simon Dentith for pointing me toward McGann's text.

47 Rossetti, 'After Death', in *Selected Poems*, pp. 33–4.

beyond her own life to set out the conditions of her lover's memory (conditions which gradually break down) that she talks herself out of existence.[48] In 'A Chilly Night', too, the dead are encountered from this side: the speaker gets up 'at the dead of night' to see an apparition of spirits floating outside her window, one of which is her mother, whose words she cannot understand. Again, this is a liminal poem, within which ghosts and live speaker enter into the same twilight zone of soul sleep, *almost* connecting there except for the absence of audible discourse:

> I strained to catch her words,
> And she strained to make me hear;
> But never a sound of words
> Fell on my straining ear.[49]

What is interesting about this poem (and this is also true of 'Maude: A Story for Girls') is its exploration of a poet's worse-case scenario: more distressing than meeting the dead is being unheard and unhearable (also, one might say, one of the potential frustrations of the analysand or the silenced hysteric). It is, then, not enough for the poem to push toward the disturbance of encountering the *revenant* (a returned dead person) – what it must also do is imagine the frustrated silence of mismeeting, the terminal mismatching of senses: she sees that her mother only *seems* 'to look at [her]', and

> She opened her mouth and spoke;
> I could not hear a word ...
> She knew I could not hear
> The message that she told

– so that finally 'Living had failed and dead had failed,/ and I was indeed alone'. In 'At Home', which I shall discuss more closely in a moment, the *revenant* speaker returns to the place in which she was once at home to find it uncannily *not* home. Rossetti's speakers demonstrate an uncanny mobility across the life/death divide which occurs in tandem with a total commitment to the transformative powers of darkness.

What is at stake in this projection into death as a waiting-room from which one looks *back*, this identification with the corpse, seeing from its point of view? Whilst undoubtedly Le Fanu is conflating at least three figures of fear in Carmilla in order to create the most disturbing possible monster – the returning dead person, the woman claiming her own space and the rapacious lesbian corrupting young women into the ways of homo- (as well as haemo-) eroticism – to many a modern reader she emerges from the tale as its most appealing character, romantically pur-

48 Rossetti, 'Remember', in *Selected Poems*, p. 35.
49 Rossetti, 'A Chilly Night', in *Selected Poems*, p. 59–61.

suing women with determined passion. Against the grain her message might be: in the land of the living women's power is limited – in the realm of the undead, which lies across and between that of the living, it might be greater. Speak as a corpse (as Rossetti does) or as a reanimated dead woman (albeit a sexually powerful one – as Carmilla does) and you might counter the negatives of living by embracing the positives of an unterritorialized zone. The post mortem thus replaces the pre-Oedipal as an alternative feminist utopia.

IV
Sympathy for the Devil

The politics of this are dubious to say the least. If horror stories must have monsters, then Carmilla is an undead monster whose desire is finally for the dissolution of the ego, and who only finds adult female 'flight' supernaturally. Rossetti's corpse-women, meanwhile, succeed only by the most extreme form of self-abnegation, as the last three lines of 'After Death' demonstrate amply: 'He did not love me living; but once dead/ He pitied me; and very sweet it is/ To know that he is warm though I am cold'. Earlier I discussed the death drive as a theory of zero excitation, but Freud also developed it as an explanation for the prevalence of masochism, which suggests that a key channel for desire is pleasure in pain, culminating in the destruction of the ego, if not the body: masochism then embraces the journey toward not-being.

In his encounters not only with sadism and masochism but with the desire radically and ecstatically to 'let go' of the self, Freud required a theoretical model which disturbed a psychoanalytic optimism based on humanistic life-instincts and healthy libidos. In *Instincts and Their Vicissitudes* (1915) he writes 'Hate, as a relation to objects, is older than love',[50] and with the development of this theory 'hate' (death instincts) in its broadest sense begins to take over from 'love' (life instincts) as the governing mental principle. Two apparently quite *separate* forms of desire – first, to enter erotically into pain and bondage (sadistically and masochistically) and second, to reach a moment of ego-extinction, the 'nirvana' of the 'zero-point' of the self (what Freud at one point calls 'primary masochism', a radical self-abnegation) – are finally explained by Freud *together* as two forms of the death drive, introduced most fully, as we have seen, in *Beyond the Pleasure Principle*, and then maintained as the basic structure of his theory of the instincts.

It is important to keep thinking of these two components together, even though this is not generally what has happened in subsequent

50 PF vol. xi, p. 137.

psychoanalytic history. Both the first and the second form of desire here are forms of return, routes backwards, toward the abolition of subjective unities which prevails before and after life. One (erotogenic sadism and masochism) prioritizes in its desire the rending and tearing of the self from its dignity and sense of control, whilst the desire of the other (that of the primary masochist) looks straight to the moment of disintegration. This primary masochism is central to desire: desire is the desire for radical non-agency, for discharge, for the dissolution of the ego – that exquisite moment of 'letting go' when the subject ceases to grasp its boundaries, ceases to be as a subject.[51]

This move from sex to death is clearly a radical one; although a number of texts written during the years around the First World War are crucial in the development of this theory, it is with *Beyond the Pleasure Principle* that the death drive is established as the central concept of Freud's later years. However, this does not stop it from sinking below the surface of subsequent, post-Freudian psychoanalytic readings and histories which often erase the traces of it altogether. From its position as the repressed in Freud's earlier writings, death is placed centre-stage during and after the First World War, only to be pushed to the margins of the picture again after his own death.

It is true that several factors, which I have traced above, led Freud to his formulation of the drive, and these as individual elements have continued to occupy important ground in psychoanalytic thought – sadism and masochism as sexual and 'moral' responses, or self-abnegation read as 'feminine' masochism. But the death drive's precarious and disavowed life after (Freud's) death has meant that it has primarily been recognized in specific discussions of aggression, pleasure in pain, or politically problematic psychic states (I am thinking of Reich's argument with Freud in the 1940s, and Fenichel's political-psychoanalytic work on the death drive, as well as recent feminist analyses of feminine masochism which pick up on late-Freud[52]). The death drive has then been read and used in two distinct ways: as a theory of aggressivity and as a

51 I discuss the differences between primary and secondary masochism in 'Submission and Reading: Feminine Masochism and Feminist Criticism', *New Formations*, Special Issue, 'Modernism/Masochism' (Spring 1989).

52 For a succinct account of Reich's position on the Freudian death drive, see Wilhelm Reich, *Reich Speaks of Freud* (Harmondsworth: Penguin, 1975). For Otto Fenichel's position on the death drive see 'Critique of the Death Instinct' (1935), in *The Collected Papers of Otto Fenichel* ed. Hanna Fenichel and David Rapaport (London: Routledge & Kegan Paul, 1954); and Russell Jacoby's *The Repression of Psychoanalysis: Otto Fenichel and the Political Freudians* (New York: Basic Books, 1983). A number of recent feminist texts have explored the question of masochism from various angles, including Paula Caplan's *The Myth of Women's Masochism* (London: Methuen, 1986), Maria Marcus's *A Taste For Pain: On Maoschism and Female Sexuality*, trans. Joan Tate (London: Souvenir, 1981), as well as Kaja Silverman's work on masochism in film theory in *Male Subjectivity at the Margins* (New York and London: Routledge 1992).

specifically determined model of return, brought into being at a historical and cultural turning-point (as well as a key moment of crisis for Freud biographically) which has importance for our readings of the moment of the First World War. As one of the key concepts of modernism, it is legacy both of Nietzsche's eternal recurrence and, in quite a different way, of Hegel's master–slave dialectic.

In his 1991 text *Death and Desire*, Richard Boothby makes the point that the death drive has only had limited currency since Freud's death. 'According to a common view,' he writes,

> we need not worry ourselves about a specifically self-annihilating drive. We need only recognize the natural tendency in human beings toward aggressivity and destruction.[53]

The point is important. For many, the death drive has become a twentieth-century recoining of essential animal aggression – a comforting and dubiously acceptable notion of natural sadism. In this context, Kaja Silverman makes the point in her essay, 'Masochism and Male Subjectivity', that some perversions are more 'perverse' than others. 'It is unfortunate but not surprising', she writes, 'that the perversion which has commandeered most of the literary and theoretical attention – sadism – is also the one which is most compatible with conventional heterosexuality.'[54] She goes on to note that,

> The first thing that Freud says about sadism in *Three Essays* is that "the sexuality of most male human beings contains an element of aggressiveness – a desire to subjugate".

Yet it was partly because the earlier point of the *Three Essays* was posed as a question – why sadism? why aggressivity? – that the death drive was required as a model of explanation. Without the death drive, aggressivity is allowed to rest as an essential animal characteristic. Laplanche reads the death drive in these terms, its importance being that it demonstrates that aggression is not primarily turned outwards against the other but (primally) toward the self. As he writes in *Problématiques IV* of psychoanalysis' refusal to acknowledge this:

> in every case, there is a refusal of the essential thesis of Freud which affirms that the death drive is in the first instance turned, not toward the outside (as aggressivity), but toward the subject, that it is radically not a drive to *murder*, but a drive to *suicide* or *to kill oneself*.[55]

53 *Death and Desire: Psychoanalytic Theory in Lacan's Return to Freud* (London and New York: Routledge, 1991), p. 9.
54 'Masochism and Male Subjectivity', p. 187.
55 *Problématiques IV: l'inconscient et le ça* (Paris: Presses Universitaires de France, 1981), p. 230, quoted by Boothby in *Death and Desire*.

Yet sadism is far more culturally (and sexually) acceptable than masochism; aggression outward is less disturbing than the drive to self-abnegation or dissolution which underpins the death drive. Laplanche underlines the point that the masochism of the death drive has been systematically repressed in favour of a more acceptable notion of aggression at the heart of psychic life, that aggressivity has been emphasized against the more fundamental, masochistic aspects of the death drive.

So why is it that sadism is a more culturally 'comfortable' notion than masochism, if not simply a personal or historical symptom (the theoretical articulation of the aggression and chaos writ large on the Western front), then the most utilizable aspect of the death drive, and the only aspect of it which has any continuing analytic 'life'? Between these more comforting readings, which place the theory either in terms of sadism or of world-historic trauma, lies another way of reading the drive, as a key cultural articulation of a masochistic aesthetic, whether this be the 'dismembering aesthetics', the self-referentiality of the modernist movement which spawned it, or of earlier texts which foreground the shrunken, liminal or undead subject, speaking back from the margins beyond.

How, then, might these differences, between primary and secondary masochism (taking the various forms of that general desire for non-agency, the longing for the abolition of the subject, or the pleasure taken in pain at the hands of a sadist) be mapped onto our texts? Both *Carmilla* and Rossetti represent *both* sensual experiences of pain which take place within a master–slave power structure, *and* they celebrate the self in its most minimal, negative form. For Rossetti, the closer she can get to non-being and still have a voice, the more she can say. Carmilla pushes her women into erotic experiences of egoistic dissolution within which pain doesn't matter – 'Love will have its sacrifices. No sacrifice without blood'[56] – and her own account of her initiation into vampirism is fed through a filter of distorted and diffuse subjectivity:

> 'I see it all, as divers see what is going on above them, through a medium, dense, rippling, but transparent. There occurred that night what has confused the picture, and made its colours faint. I was all but assassinated in my bed . . .'[57]

In response, Carmilla is now the assassin, and Laura's bed is the site of a 'vague and strange' slow sexual murder, which feels no worse than the sensation 'of that pleasant, peculiar cold thrill which we feel in bathing, when we move against the current of a river'.[58] Carmilla commands with the force of a Sadeian libertine: 'You must come with me, loving me, to death; or else hate me, and still come with me, and *hating* me through

56 'Carmilla', p. 101.
57 Ibid.
58 p. 105.

death and after'.[59] Earlier, she claims a dissolution of the boundaries between their identities, as her control over Laura (expressed as her *own* 'humiliation') increases in tandem with her thirst:

> 'In the rapture of my enormous humiliation I live in your warm life, and you shall die – die, sweetly die – into mine. I cannot help it; as I draw near to you, you, in your turn, will draw near to others, and learn the raptures of cruelty, which yet is love.'[60]

Like the victim who goes on to perpetrate and replicate identical crimes as those committed against her, the vampire is caught in a sado-masochistic tangle. If both Rossetti's self-abnegation and ignorance of boundaries and Carmilla's avaricious sadomasochism are vampiric, then on which side of the master–slave divide do we situate it? Carmilla calls 'the irresistible law' of vampirism her 'strength *and* weakness'.[61] One of the key moments in *Dracula* highlights this problem, which is also at the heart of the psychoanalytic account of the perverse subject:

> With his left hand he held both Mrs Harker's hands, keeping them away with her arms at full tension; his right hand gripped her by the back of the neck, forcing her face down on his bosom. Her white nightdress was smeared with blood, and a thin stream trickled down the man's bare breast, which was shown by his torn open dress. The attitude of the two had a terrible resemblance to a child forcing a kitten's nose into a saucer of milk to compel it to drink.[62]

This is an extraordinary scene, within which both parties are both victim and violator: Dracula forces Mina to prey upon him, whilst Mina is resistant but still drinks, a substance which is likened to milk from a *man's* breast – milk is not only what the kitten really wants but also what might be good for it. Dracula is the child here, and it is *his* bodice which has been ripped, whilst Mina's nightdress is stained, not with her own blood but with his. He is the bodily injured party here, yet he is also the demonic, active presence. Perhaps despite themselves, both Le Fanu and Stoker suggest that vampirism muddies the waters of any clear-cut reading of gendered power-relations and identification,[63] since it betrays the possibility that pleasure is dangerous, and our victims might also be violators.

59 p. 100.
60 p. 89.
61 Ibid.
62 *Dracula*, p. 336.
63 These are also texts which, in their very ideological confusion, open themselves up to potential readerly pleasures which challenge author- and gender-based models of identification. For Andrea Weiss, the text's lesbianism is only finally appropriated into a patriarchal fantasy once it is reworked cinematically as the 1970 film, *The Vampire Lovers*, whilst the novel itself has, for Gene Damon, 'long been a sub-basement Lesbian classic'. Against this preference for the textual, Bertha Harris cites 'Christopher Lee, in drag, in the Hammer Films, middle period' as 'my ideal lesbian' (both quotations from Weiss, 'The Vampire Lovers', in *Vampires and Violets: Lesbians in the Cinema* (London: Jonathan Cape, 1992)).

V
Home Is Where the Uncanny Is

If we look at these texts from a rather different point of view, however, what both Rossetti and Carmilla are doing is privileging entirely the wrong side, prioritizing one side of a binarity, but both choosing the negative option. Carmilla reads death from the perspective of death – 'this disease that invades the country is natural. Nature. All things proceed from Nature – don't they?'[64] Again I am reminded of David Cronenberg, who has also said that 'a virus is only doing its job' and, 'to understand it from the disease's point of view, it's just a matter of life. I think most diseases would be shocked to be considered diseases at all.' Freud focuses on this too when he writes that it is the 'goal' of all life to return to the 'natural' world of the inorganic, and that life is only a circuitous route back to quiescence in the grave. We come from nature and, through the benign influence of disease, we return to it. Freudian instincts are essentially conservative, especially the death drive as primary instinct. So the problem of how to *defer* arriving at the moment that the drive *wants* is acute – by taking (in Roger Ascham's terms) the 'long wandering' route rather than a short cut the life-instincts can defer the 'conservative' aim of the death drive for as long as possible. As Freud gloomily writes to Fliess in 1888, 'life is generally known to be very difficult and very complicated and, as we say in Vienna, there are many roads to the Central Cemetery'[65] – the problem is how to travel by the longest one.[66]

Carmilla, as a classic Victorian horror story, also stands as an exemplary uncanny text (you can almost tick off the motifs from Freud's essay of that name): the repetition or anagrammatical repetition of names (Carmilla/ Millarca/ Marcia/ Mircalla); the doubling of dreams ostensibly experienced by two different people and the double image of Carmilla in body and picture (she arrives at the house a stranger, and it is discovered that she mirrors a centuries-old painting already hanging there, finding her *doppelgänger* in an image); and finally the repetition of narrative (Laura's account of her own demise prefiguring the General's account of his daughter's). But the uncanny is also characterized for Freud by the *déjà vu* of homecoming, returning or coming to a place

64 'Carmilla', p. 95.

65 Freud, letter to Fliess, in *Complete Letters of Sigmund Freud to Wilhelm Fliess*, p. 22.

66 For Peter Brooks, this could then mean maximizing your sub-plots: 'the subplot stands is one means of warding off the danger of short-circuit, assuring that the main plot will continue through to the right end .... Deviance, *détour*, an intention which is irritation: these are characteristics of the narratable, of "life" as it is the material of narrative, of *fabula* become *sjuzet*. Plot is a kind of arabesque or squiggle toward the end' ('Freud's Masterplot', p. 292). Later he writes: 'The story of Scheherezade is doubtless the story of stories' (p. 299).

which is horribly but unfixably familiar. As Freud writes in his 1919 essay 'The "Uncanny"':

> It often happens that neurotic men declare that they feel there is something uncanny about the female genital organs. This *unheimlich* place, however, is the entrance to the former *Heim* [home] of all human beings, to the place where each one of us lived once upon a time and in the beginning. There is a joke saying that 'Love is home-sickness'; and whenever a man dreams of a place or a country and says to himself, while he is dreaming: 'this place is familiar to me, I've been here before', we may interpret the place as being his mother's genitals or her body.[67]

Home is finally the place from which one comes – the mother or, more specifically the woman's genitals, and the 'home' or origin of the inorganic. In this last sense, by the time Freud develops the death drive, home is also where we return *to*. Death is a woman's – the mother's – body.[68]

This is also an image which figures centrally in *Carmilla*, with its emphasis on a death-bringing oral relationship between two women (Rossetti's 'Goblin Market' features something of this conflation, of a culminating moment of feeding from a woman's body which acts as a path through to salvation). But there is a ghoulish feminism in Carmilla's preference for death. Death promises a return to the mother, and an act of female solidarity. If death, for the Freud of the 1920s, is a final 'homecoming', then Carmilla prefigures this, saying, 'Why, *you* must die – *everyone* must die, and all are happier when they do. Come home.' Throughout the text Carmilla is set up as a kind of alpha and omega figure: Laura's earliest memory is of a dream of Carmilla, and though dead she is also present in the closing lines – 'often from a reverie I have started, fancying I heard the light step of Carmilla at the drawing-room door'.[69] Laura also dreams that her mother warns her 'to beware of the assassin', which is accompanied by an image of Carmilla 'in her white nightdress, bathed, from her chin to her feet, in one great stain of blood'.[70] Here it is unclear whether Carmilla is good mother or bad, victim or assassin, the subject or object of the warning.

It is not surprising then that these two ideas – of the mother's body, and of death, as 'home' – should be conflated by psychoanalysis. With characteristic ethnocentricity Freud writes in 'The "Uncanny"':

67 Freud, 'The "Uncanny"' (1919), PF vol. xiv, p. 368.
68 There is also an incipient national discourse at work here, as Denise Riley has pointed out to me. Nationalism would then be read as the death drive writ large.
69 p. 137.
70 p. 106.

> Since almost all of us still think as savages do on this topic, it is no matter for surprise that the primitive fear of the dead is still so strong within us.[71]

The *revenant* is consequently interpreted almost exclusively as an ancestor or parent returning from the repressed. Ernest Jones in particular identifies the *revenant* as 'the spirits of deceased parents . . . The attitude of awe and fear in respect of dream visitors from the dead has been thought to be one of the main sources of ancestor-worship'.[72] The significance of this 'homecoming' for feminism is, however, tangential to this discussion of the gothic and the death drive. Death, and problems involved in this fabled female response to it, might give us another way of exploring the more immediate feminist celebration of the pre-Oedipal, that 'space outside' which is subject to a similar range of political problems but which might be more clearly exposed by exploration of what is at stake in deifying, on the behalf of women, that impossible space at the *other* end of life, undead zones which are alternative escape-routes: Rossettian liminality or soul sleep, and Carmilla's post-Symbolic. It is not the role of this chapter to enter into a critique of pre-Oedipal feminism through a reading of post-mortem feminism; readers of the first might, however, look to the second as its literary antecendent.

I want to conclude this chapter by reading another curious liminal poem by Rossetti, 'At Home', as well as by taking a closer look at an essay I have thus far mentioned several times but never explored fully. Freud's 'The "Uncanny"', is perhaps the most readily applied work of his later phase, offering a literary reading of repetition compulsion and 'what is frightening . . . what arouses dread and horror . . . repulsion and distress'[73] through a partial reading of Hoffman's story, 'The Sand-Man', and the disturbing images and emotions which erupt in and through it. The broad sweep of the piece takes the notion of an 'uncanny' experience through a series of conceptual reversals. First of all, Freud discusses the German word *unheimlich* (uncanny – literally 'unhomely') as

> obviously the opposite of *'heimlich'* ['homely'], *'heimisch'* ['native'] – the opposite of what is familiar; and we are tempted to conclude that what is 'uncanny' is frightening precisely because it is not known and familiar.[74]

He then moves through a long series of definitions and etymologies 'to find that among [the] different shades of meaning [of the uncanny] the word *'heimlich'* exhibits one which is identical with its opposite, 'unheimlich' . . . on the one hand [the word *'heimlich'*] means what is familiar and agreeable, and on the other, what is concealed and kept out of sight':[75]

71 Freud, 'The "Uncanny" ', p. 365.
72 Jones, *On the Nightmare*, p. 63.
73 pp. 339–40.
74 p. 341.
75 p. 345.

> everything is *unheimlich* that ought to have remained secret and hidden but has come to light. . . .
>
> Thus *heimlich* is a word the meaning of which develops in the direction of ambivalence, until it finally coincides with its opposite, *unheimlich. Unheimlich* is in some way or other a sub-species of *heimlich.*[76]

Much of the essay is itself concerned with the various qualities of the uncanny and their significance in mental life, as Freud explores point by point the conditions under which the uncanny is activated. Through Hoffmann's story he is able to identify a number of motifs which have peculiar significance to the post-death drive subject, and which only come to the fore in Freud's work at this moment after the First World War. His primary concern is 'selecting those themes of uncanniness which are most prominent, and with seeing whether they too can fairly be traced back to infantile sources';[77] so, having identified the key motifs of the literary uncanny, Freud then accounts for their power with reference to the development of the individual. The uncanny 'is in reality nothing new or alien, but something which is familiar and old-established in the mind and which has become alienated from it only through the processes of repression';[78] 'the uncanny [*unheimlich*] is something which is secretly familiar [*heimlich-heimisch*], which has undergone repression and then returned from it'.[79]

Without wishing to decode these motifs into a specific account of Rossetti as 'origin' of the structures of repression which give rise to her poetic uncanniness, it must be said that Freud's concerns here are themselves a remarkable (not to say uncanny) echo of Rossetti's (as well as those of *Carmilla*). The power of the double or *doppelgänger*, the 'doubling, dividing and interchanging of the self', 'the constant recurrence of the same thing – the repetition of the same features or character-traits or vicissitudes, of the same crimes, or even the same names through several consecutive generations',[80] each has its incarnation in the literary texts under scrutiny here. Indeed, many of these images return to the dead body or even the reanimated corpse; 'the "double"', writes Freud (noting a debt to Otto Rank), 'was originally an insurance against the destruction of the ego . . . and probably the "immortal" soul was the first "double" of the body'.[81] Thus the voice of 'After Death' becomes that of the double of the 'I' who also lies dead beneath the shroud, hovering over and haunting the scene which it also corporeally inhabits below. Similarly, the spirit-moth of 'Two Thoughts of Death' emerges as the

76 pp. 345 and 347.
77 p. 356.
78 pp. 363–4.
79 p. 368.
80 p. 356.
81 Ibid.

double of the decaying corpse. Both poems are, then, fantasies which guarantee the survival of the undead ego whilst also encountering the return of the body to the inorganic. Rossetti's corpse-poems, with their central reliance on a spiritualized form of '[a]pparent death and the re-animation of the dead',[82] also offer what Freud calls 'a particularly favourable condition for awakening uncanny feelings . . . when there is intellectual uncertainty whether an object is alive or not, and when an inanimate object becomes too much like an animate one'.[83]

But it is 'At Home' which addresses the central concern of the uncanny: repetition compulsion and the ambiguity of *heimlich* as precisely *unheimlich*. Here, the dead speaker who comes home is *anything but* at home. The poem also plays out an important vampiric point, that the problem of 'undeath' lies not in appetite and endless orality but in the prospect that release will never be reached – 'At Home' suggests the awful possibility that death condemns the subject only to a sequence of returning, to a 'before' from which you are for ever exiled. 'When I was dead', she writes, 'my spirit turned/ To seek the much-frequented house':

> I passed the door, and saw my friends
>   Feasting beneath green orange-boughs;
> From hand to hand they pushed the wine,
>   They sucked the pulp of plum and peach;
> They sang, they jested, and they laughed,
>   For each was loved of each.[84]

What she then discovers is that all that the living are doing is looking forward to a 'To-morrow' which will never come:

> 'To-morrow shall be like
> To-day, but much more sweet.'
>
> 'To-morrow,' said they, strong with hope,
>   And dwelt upon the pleasant way:
> 'To-morrow,' cried they one and all,
>   While no-one spoke of yesterday.
> Their life stood full at blessed noon;
>   I, only I, has passed away:
>   'To-morrow and to-day', they cried;
> I was of yesterday.
>
> I shivered comfortless, but cast,
>   No chill across the tablecloth;
> I all-forgotten shivered, sad
>   To stay and yet to part how loth:
> I passed from the familiar room,

82 p. 369.
83 p. 354.
84 *Selected Poems*, pp. 76-7

I who from love had passed away,
Like the remembrance of a guest
That tarrieth but a day.

The point is not only that she has been forgotten, but that the tomorrow toward which the living are looking is actually the place in which she now resides. The better place ('much more sweet') is only the cursed experience of eternally revisiting the 'home' from whence you came and to which you no longer belong, as existence turns itself in closed circles which do not even offer the hope of arrival in the form of extinction. Instead, having left home in the most radical possible way, the voice is forced into a trajectory back to the now defamiliarized familiar, which she must revisit as a stranger. There is no arrival, no end to it. The only hope of escape from this modernistic wandering lies in Rossetti's use of the past tense: the poem opens 'When I *was* dead', suggesting that this is a phase which is over, that the circle is *not* closed.

Vampires also never die – this is what is disturbing about them, they are infected by what Van Helsing in *Dracula* calls 'the curse of immortality'. Eternally committed to the liminality of the undead state, a soul sleep from which one never awakens, vampires act out a myth of the death drive. For just as vampires never die, so the subject of the death drive as such never experiences reaching its goal. The death drive thus becomes the drive *par excellence*, in that it is entirely and explicitly predicated on *lack*. Unlike, say, forms of sexuality (for which a love-object will be designated), the death drive has no achieved goal, it is fundamentally objectless; when the goal is reached, there is no subject. Thus, unlike other drives which can fantasize toward some sort of arrival in satisfaction, the death drive is only ever understandable or experienceable in its 'driveness'. It is a drive which can only be 'present' in its *flight towards* rather than in its *arrival in* its goal. That vampires should fly is thus highly appropriate for their figuration of the drive. Driving toward the goal rather than arriving in it is the only way we can characterize the death drive, and in this sense too it epitomizes the drive *per se*. Carmilla might advocate vampiric flight as a better place, but she too is looking for some sort of arrival which she only reaches at that characteristic moment of phallic cleansing when the stake is driven in. The vampire never arrives: objectless, unsatisfied, the undead, like the living, do not rest.

VI
Modernism and Masochism

The connections with the death drive here are multiple. 'At Home' loops the thread of life and death back on itself as neatly as if it were an exercise in the poetics of *Beyond the Pleasure Principle*. Clearly this is a prime text of the uncanny, from the title itself right through to its resounding point that home is precisely not-home. Yet the trajectory of the 'I's' wandering moves the poem toward a more modernistic aesthetic than that which usually characterizes Rossetti's work. I have already pointed towards the possible masochism of the modernism which was the context of Freud's theory of the death drive. But something else is going on in the movement of this poem which underlines a returning trajectory in poem, theory and the movement itself.

Whilst this discussion has primarily focused on the nineteenth-century gothic, one of its principal concerns has been to trace a kind of circularity in these texts: projection forward as a form of return. Rossetti slips beyond only to look back; Carmilla's death – a better place which is also the mother's space – inaugurates only a compulsive hunger for the living. There is a modern circularity about these projections. What, then, are the implications of the death drive for our understanding of the self-refentiality of cultural forms which also keep returning to and interrogating themselves? This might be to stress elements of modernism which are masochistic or narcissistic, looking in rather than out: tied into circles of return and self-reference which are more characteristic of masochism than sadism. Sadism looks to another, outside itself, for its satisfaction: the sadist is locked into a relationship of 'power-over' which flings her thoroughly forward into the world.

Notwithstanding the wandering, exogamic movement which is often taken to epitomize modernism – the modern subject in his Lukácsian 'metaphysical homelessness' leaving, moving out and on, rootless – these patterns of repetition and of a masochism which turns in its desire back to itself, articulate a centripetal rather than a centrifugal desire: to return, repeat, come back into the moment of origin, which itself disappears into myth once the circle closes. Against a wandering modernism is pitched a circular modernism, which is about homecoming as much as it is about homelessness; but home is no longer what you think it is – after Freud's uncanny, 'home' is anything but 'homely'. Alongside the desire for exile, the desire to fling oneself beyond the territory of birth and origin, is placed this desire for return. Exodus is countered by incest, narcissism and masochism as various forms of return to the self, the familiar, the end which becomes the start.

This pattern back to the self and then back to the moment before the self, to an 'original' state of blank obscurity, is articulated in various ways in both Nietzsche and Freud. Although my concern is with the death drive, I am also thinking not only of Nietzsche's Eternal Recurrence but of Heidegger's reading of its genesis in *Thus Spake Zarathustra*. Zarathustra 'enjoys' or befriends his solitude (*'genoß er seines Geistes und seiner Einsamkeit'*[85]), a kind of narcissism of thought which, as

85 Friedrich Nietzsche, *Also Sprach Zarathustra*, in *Werke*, vol. ii, ed. Karl Schlechta (Munich: 1981), p. 551.

Heidegger points out, enables the thought of the circle, the fully formulated theory of Eternal Recurrence which is to come. Thus Zarathustra converses with his soul: 'the essence of thinking resides in the soul's solitary conversation with itself', writes Heidegger,

> the telling self-gathering which the soul itself undergoes on its way to itself, within the scope of whatever it is looking at.
>
> In converse with his soul Zarathustra thinks, his "most abysmal thought"...[86]

The eternal recurrence thus is not just *what* Zarathustra eventually conceives, it is *how* he thinks it: in this motion of self-return he finds the model of and for historical return. Something of this is going on in Freud too, and this is just one of the ways in which Freud and Nietzsche uncannily repeat each other on the issue of repetition. If returning and turning inward is both the pattern and result of Zarathustra's meditation, what happens with Freud in his development of the death drive is initially a turning-back to his medical origins at the same time as he sets up the idea of return itself: '[E]veryone owes nature a death and must expect to pay the debt.' Life is lived on borrowed time; return to the point of loan and nature will reclaim from you that which you took at the start. Yet what is interesting here is not just what Freud is saying, but that he is saying it in terms of a revisitation of earlier positions. On the issue of death itself, Freud is also handing something back to the strictly natural science which he abandoned in the 1890s when he left positivistic medicine in pursuit of a more seductive form of therapy. In handing back the body he begins epistemologically to make amends. Just as 'everyone owes nature', so he 'owes' natural science the finally unanalysable thing – the returning-to-the-inorganic dead body itself – once analysis has become most violently 'terminable'.

To mitigate the Heideggerian sense here that this is the meditation of a whole soul, turning into itself in order to reinforce its unified selfhood, it must be stressed that what the self finds here is profoundly disruptive. In 'coming home' it is doing one of two things: either it is dying (home as death) or it is experiencing home as *unheimlich*, discomforting and difficult, alien. Similarly, Nietzsche's eternal recurrence is established in tandem with a theory of subjectivity which dismembers the self rather than enshrines it, which problematizes identity in terms which state that the ego is nothing more than a grammatical structure or habit. In both philosophies, what is 'home', the most familiar, is radically defamiliarized.

86 Martin Heidegger, *Nietzsche*, vol. ii: *The Eternal Recurrence of the Same* (San Francisco, CA: 1984), p. 218.

In a chapter so concerned with circles I want to turn this ending back to the beginning, to find Tess still looking in her mirror, realizing that the date at the heart of her life is that of 'her terminus'. In such a moment of self-gathering, the self-in-the-mirror tells her that for all her nomadism she is finding 'a short way by a long wandering', forward and back to the spot from which her mother's, rather than her father's family come: in her final place, just as she is about to die, she also uncannily returns to the mother, remembering that 'One of my mother's people was a shepherd hereabouts . . . So now I am at home.'[87]

[87] Hardy, *Tess of the D'Urbervilles*, p. 484.

6

The Resistances to Analysis: *Jude the Obscure*

This final chapter is about failure. Or rather it is about the *uses* of failure. There have been moments in this book when I have tried to highlight the fact that, for psychoanalysis, failure is not necessarily a problem; upon a gap, a fissure, a point of irreconcilability in the self, our incomplete identities rest. That we fail to be whole underpins what is interesting about the psychoanalytic model of the subject (even though, as a therapeutic practice, psychoanalysis strives to minimize the debilitating effects of that 'failure'). This also accounts for much of what is at stake in, and what is compelling about, certain cultural constructions of the subject in writing and film (as well as other art forms I have not been able to discuss here). 'Failure', then, is not a problem but a condition of subjectivity.[1]

But 'failure' for psychoanalysis might also be about something which takes place during the analysis itself. Analysands *resist*, their discourse breaks down, they fall into silence, they say 'No', and these are moments which are integral to the process of analysis, which reads some kinds of failure as forms of success. Resistance, negation, disavowal are responses which Freud overtly discussed, and in this chapter I shall look briefly at how these are psychoanalytically read in reverse. Rather than being interpreted as the responses of absolute defeat, these moments point to a different kind of failure, the failure of repression – the juncture at which the seal begins to break. Yet there are real losses too, which psychoanalysis cannot account for or appropriate – the negative therapeutic reaction, the interminable analysis, for instance. Analysands (like Dora) say 'No'

[1] Once again, the point made by Jacqueline Rose, and quoted in Chapter 1, has a purchase here: 'The unconscious constantly reveals the "failure" of identity . . . "failure" is something endlessly repeated and relived moment by moment throughout our individual histories' (*Sexuality in the Field of Vision*, p. 91).

to the process of analysis itself, in a way which Freud cannot fully account for – this is a refusal which cannot easily be assimilated into a theory of a positively incomplete identity. I have worked with a number of psychoanalytic models and theorists in this book, primarily to highlight as accessibly as possible their integral relationship with some appropriate literary texts. But just as Dora refuses and resists, then so do texts. What happens, then, when the text says 'No'? How can psychoanalysis address a text which is *in*appropriate, or which manifestly refuses to be appropriated?

One such text is Thomas Hardy's *Jude the Obscure* (1896). This is the story of Jude Fawley, a humble, self-taught stonemason, whose thwarted attempt to enter the world of knowledge at Christminster (Oxford) is woven together with his difficult relationships with women, first with his earthy wife Arabella and then with his intellectual lover (and cousin) Sue Bridehead. Whilst the novel overtly deals with problematic sexual relationships, divorce and the possibilities of what were then revolutionary family structures (Jude and Sue live together unmarried, and have 'illegitimate' children), it is canonically central for other reasons: its humanism (through the unfolding tragedy of Jude's self-perception), its analysis of the individual's relationship to the class structure, and its discussion of educational questions (Jude's relationship with formal learning and with the University which excludes him even as he – as a stonemason – rebuilds its walls[2]). Furthermore, discussions of gender in the novel have focused on how it is situated in terms of the Victorian 'Woman Question', specifically on the figure of Sue as a 'New Woman', who attempts to lead something of a bohemian life, earning her own living and (for part of the novel at least) embracing unconventional, free relationships. Precisely because it has been read so successfully by humanist and Marxist criticism, *Jude the Obscure* offers something of a challenge to the psychoanalytic reader.

I

Jude the Obscure and Sue the Articulate

> A perception of the failure of things to be what they are meant to be, lends them, in place of the intended interest, a new and greater interest of an unintended kind.[3]
>
> Thomas Hardy, *Journal*, 1st January 1879

2 In his pithy Introduction to the Macmillan edition of *Jude the Obscure*, which discusses the novel as a tale of fatalistic false consciousness produced by a 'false society', Terry Eagleton writes: 'Jude's labour-power is exploited literally to prop up the structures which exclude him'. (Thomas Hardy, *Jude the Obscure* (London and Basingstoke: Macmillan, 1978), p. 13).

3 Thomas Hardy, *Journal*, 1 Jan. 1879.

If we find access to this text through the channels of psychoanalysis initially difficult, this may be an effect of the fact that the text itself is also about the inaccessible. It is a story of denial – a whole system of denials. Jude is excluded, left out, *shut* out, and if this is partly because of his educational and social position, it is also because of his sexuality. He is denied access, he is *resisted*, and by no one as much as the woman he loves. In a resistant text, it is Sue who resists most. Six years before Dora left Freud, the publication of *Jude the Obscure* offered readers a number of similar leave-takings. Repeatedly, Sue turns her back, withdraws in a hurry, or moves away before she says too much, leaving Jude standing on the station platform without any clear sense of where she's gone or why. She leaves Jude, over and over again, but she also leaves us, in the sense that her persona builds a wall against critical delineation or intrusion. Whilst we could not easily figure Jude as the text's diegetic analyst (I shall come on to his perplexity in a moment), we should pause on what is at stake in Sue's resistance.

Sue is a problem; she is 'something of a riddle'.[4] She does not simply block understanding by turning her back, but by offering – as if to explain everything – a handful of pieces which seem to come from different jigsaws. She is the construction of contradictory discourses, or a discourse of contradiction. Analysis of one short section, chosen fairly randomly, bears this out. Sue has run away to Jude's lodgings and, her own clothes being wet, 'masquerades' as him by changing into his Sunday suit (becoming a 'better' version of him). The whole episode is peppered with contrary remarks as Sue discusses her identity but says precisely nothing which can fix her to anything, all of the time dressed in a man's clothing. 'I am not particularly innocent,' she says first, but 'with an ostensible sneer, though he could hear that she was brimming with tears'. Then she says, 'People say I must be cold – sexless', and a moment later, speaks 'in a voice of such extraordinary tenderness that it hardly seemed to come from the same woman who had just told her story so lightly'. Finally, a few lines later, Jude 'felt that she was treating him cruelly' and '[h]er very helplessness semed to make her so much stronger than he'. Each switch takes place in rapid succession, and this all happens in the space of a page. It is no wonder that in the middle of all this, 'Jude felt much depressed; she seemed to get further and further away from him with her strange and curious unconsciousness of gender.'[5]

This is no active resistance – not yet, anyway. At worse, this is evasion, a process of refusal which acts on Jude like a linguistic dance of the seven veils. I have said that the novel is shaped by images of being shut out, but this question of knowledge denied which the novel addresses as an *educational* issue continues to be posed through Sue's denial of Jude's need

4 Ibid., 153.
5 All quotations, *Jude*, p. 169.

to know *her*. One kind of search for knowledge in the text is replaced by another: when Jude cannot gain access to formal education, he moves on to not getting access to Sue.[6] Sue's uncanny femininity thus takes over as the text's key epistemological puzzle.

This positions Jude as Sue's reader-in-the-text, and from his perplexity in 'the elusiveness of her curious double nature'[7] other readers have taken their cue. For Christine Brooke-Rose the text is characterized by a dialectic of disclosure and reserve, of 'the hidden and the revealed . . . a dialectic as cockteasing'.[8] For T. R. Wright, Hardy 'is fertile in . . . circumlocution . . . as a mode of literary foreplay',[9] and perhaps never more so than with Sue Bridehead. It is this sense of constant disappointment, of – in the words of the epigraph from Hardy's *Journal* above – failure to be 'what she is meant to be', which 'lends her a new and greater interest'. Sue's power and interest are maintained by virtue of how she is concealed rather than what she reveals. It is not simply that (as we shall see) she speaks a discourse of negatives which means that on a very fundamental level she is 'not there'. She is a talkative cipher; her 'absence' is precisely an effect of how much she says – the more she says, the less she's 'there'. Deferring the moment of arrival – sexual consummation for Jude, readerly consummation for us – Sue becomes more and more remote as the novel progresses. As Mary Jacobus puts it, it is Sue who is 'the obscure', not Jude.[10]

It is exactly this obscurity which has provoked a very particular critical response. Sue says at one point that her 'life has been entirely shaped by what people call a peculiarity in me'.[11] This is true not only of her life as the text offers it but of the critical life she has 'led' in subsequent readers' accounts of the text. Her perversities and indefinitenesses are exactly what has enticed critics into the most basic forms of analysis (for instance, Rosemarie Morgan's absurd psycho-character analysis in *Women and Sexuality in the Novels of Thomas Hardy*[12]). Indeed, the history of Hardy scholarship is marked by uneasy critical attempts to analyse

6 As Christine Brooke-Rose writes, 'Sue replaces . . . his single-minded project', 'Ill Wit and Sick Tragedy: *Jude the Obscure*', in *Alternative Hardy*, ed. Lance Butler (London and Basingstoke: Macmillan, 1989), p. 28.

7 *Jude*, p. 229.

8 p. 35.

9 *Hardy and the Erotic* (London and Basingstoke: Macmillan, 1989), p. 13.

10 See Mary Jacobus, 'Sue the Obscure', *Essays in Criticism*, 25 (July 1975), pp. 304–28.

11 *Jude*, p. 167.

12 Although this text does contain some interesting readings, Morgan still falls into the psychobiographical trap of projecting onto characters past histories and unconscious lives which are never there in print, a kind of psychoanalytic version of the 'How many children had Lady Macbeth?' question: 'if Sue fears her own sexuality this probably originated in her infancy in being taught to hate her mother and in identifying with the father who both hates the mother and rejects the mother's daughter'; *Women and Sexuality in the Novels of Thomas Hardy* (London and New York: Routledge, 1988), p. 128.

her, decipher her, pin her down and render her meaning explicit, precisely because it isn't. Following on from Hardy's clear dislike of his heroine (in his 1912 Preface he calls her an 'intellectualized, emancipated bundle of nerves'), the later critical diagnoses are not polite. Damning Dora with faint praise, Freud disdainfully notes her 'really remarkable achievements in the direction of intolerable behaviour',[13] and similarly in *Jude the Obscure* words which are ordinarily neutral or positive become negative when used in the context of Sue Bridehead. The 'bundle of nerves' also has a 'nervous little face',[14] 'she's of a thoughtful, quivering, tender nature';[15] and Jude's aunt says of her,

> I never cared much about her. A pert little thing, that's what she was too often, with her tight-strained nerves. Many's the time I've smacked her.

Words like 'nervous', 'talkative' and 'fidgety' abound – she so excessively manifests these otherwise innocuous qualities that soon her sensitivity emerges as pathological. Indeed, such a diagnosis is present even on the surface of the text, when Arabella deems Sue 'hysterical'.[16]

D. H. Lawrence takes his cue from these judgements inside the text to make the same point from the outside, engaging in a sexual diagnosis of Sue's malaise in his 1914 'Study of Thomas Hardy'. Like his later bitter critiques of Freudianism in *Fantasia of the Unconscious* and *Psychoanalysis and the Unconscious*, this reading is predicated upon an absolute belief in the primacy of separate sexual identities as the fundamental guarantor of authentic sexual health. For Lawrence, Sue does violence to the female parts of herself in a celebration of the cerebral:

> She was born with the vital female atrophied in her: she was almost male. Her will was male . . . That which was female in her she wanted to consume . . . in the fire of understanding, of giving utterance.[17]

Femininity is then 'consumed' in speech, in knowledge, in articulacy: paradoxically, the very image of unknowability in the text is obscure precisely because it – she – is so motivated by this need to *know*, and to *speak* her knowledge. Sue's articulacy emerges as her prime problem. In his important essay 'Sue Bridehead and the New Woman', John Goode writes of Lawrence's response:

13 Also quoted and discussed by Steven Marcus, *In Dora's Case*, p. 90.
14 p. 122.
15 p. 131.
16 p. 376.
17 'Study of Thomas Hardy', in *A Selection from Phoenix*, ed. A. A. H. Inglis (Harmondsworth: Penguin, 1979), p. 236.

> What is unforgivable about Sue is her utterance, her subjecting of experience to the trials of language. Lawrence, underneath the hysterical ideology, seems very acute to me, for he recognises that Sue is destructive because she utters herself – whereas in the ideology of sexism, the woman is an image to be uttered. That is to say, woman achieves her womanliness at the point at which she is silent, and therefore can be inserted as 'love' into the world of learning and labour; or rather, in Lawrence's own terms, as the 'Law' which silences all questions.[18]

Her active absence on the page only encourages readers to 'fill her up', to solve the problem of her. For John Goode this can primarily be interpreted as a sign that Sue 'alone' has little identity – she exists only as a function of Jude's desire to understand what he experiences. Sue doesn't give much away about herself because there isn't much to give; she is only there to give Jude a voice, as his verbal mirror. Sue speaks words which shape Jude; she is the agent of his articulation or, as Goode puts it, 'we don't ever ask what is happening to Sue; because it is rather a question of Sue happening to Jude'.[19]

II
Just Say No

> 'You called me a creature of civilization, or something, didn't you?' she said, breaking a silence. 'It was very odd you should have done that. . . . it is provokingly wrong. I am a sort of negation of it.'[20]

Sue talks – she is, it seems, *too* articulate. But whilst she says a lot about everyone else, she can only say 'No' about herself. For Lawrence 'The letter killeth': Sue's intimacy with the Word, her articulacy, coupled with her inability finally to reveal herself, is what condemns her. Arabella, on the other hand, offers a very different image of femininity. As Hardy writes, 'She was a complete and substantial female animal – no more, no less'.[21] In that Sue marks a difference within herself between her womanliness and the conscious processes of utterance and knowledge, she undoes herself as a complete woman. It is consequently not silence which opens up and reveals the gap here (as with the hysteric in Tania

18 'Sue Bridehead and the New Woman', in *Women Writing and Writing about Women*, ed. Mary Jacobus (London: Croom Helm, 1979), p. 101.
19 Ibid., p. 104.
20 *Jude*, p. 167.
21 *Jude*, p. 59 – although it is true that the interest of Hardy's texts is always incompleteness rather than wholeness: 'It is the *incompleteness* that is loved, when love is sterling and true . . . A man sees the Diana or Venus in his Beloved, but what he loves is the *difference*' (*Journal*, 1891; my italics).

Modleski's analysis of melodrama), but speech. For Sue Bridehead, speech reveals lack.

I have argued that her resistance, her unavailability, is exactly the thing about her which has whetted critical appetites. She does more than evade, however: she repeatedly says 'No'. In her inability to address herself and her disturbing ability to address all else, in her compulsive critique of her surroundings, her need to say 'No' to everything, she describes a psychic and social impossibility, a repeated failure of identity. I am reminded of a point Parveen Adams makes in her short paper on psychoanalysis and feminism from which I also quoted in Chapter 3, 'What is a Woman? Some Psychoanalytical Dimensions', where she writes, 'In trying to answer the question 'What is a woman?' we seem to come up against the question 'What is a perversion?'[22]

So what happens when a woman says 'No'? And is Sue's negativity entirely perverse? However clear she is about the fact that she is misunderstood, Sue is nevertheless unable to make any more positive statement about what she *is*. If any form of self-definition emerges, it is only through a web of negatives. If she says no to the world, however, it is only because of the law which has already said 'No' ('"You shan't!"') to *her*: 'First it said, "You shan't learn!" Then it said, "You shan't labour!" Now it says, "You shan't love!"'.[23] In response, she negates herself. Whereas, in Hardy's earlier *Tess of the D'Urbervilles*, Angel Clare had said to Tess, 'The woman I have been loving is not you', it is Sue who in effect says to Jude and Phillotson, 'The woman you have been loving is not me':

> '. . . you mistake me! I am very much the reverse of what you say so cruelly . . . But you are so straightforward, Jude, that you can't understand me!'[24]

Here, again, she speaks after her marriage to Jude's old teacher Phillotson:

> 'I am not really Mrs Richard Phillotson, but a woman tossed about, all alone, with aberrant passions, and unaccountable antipathies . . .'[25]

Earlier still in the novel, she resists external definition but does not compensate for this by offering her companions other forms of access to her. Denying the charge of scepticism and 'cleverness', she says,

22 See Adams, p. 41.
23 p. 357.
24 p. 225.
25 p. 226.

> 'No, Mr Phillotson, I am not – altogether! I hate to be what is called a clever girl – there are too many of that sort now! . . . I only meant – I don't know what I meant – except that it was what you don't understand!'[26]

These, then, are Sue's self-representations; she is 'not really' this, 'very much the reverse' of that, and her emotions are aberrant or unaccountable. She is simply 'what you don't understand'. 'A negation', Jude says to her at one point, 'is profound talking.'[27]

How, then, can psychoanalysis or feminism (or both) engage with this mutual articulacy and silence, with a woman's passionate ability in speaking up *for* herself which coexists, or perhaps causes, her inability to speak *about* herself? As simultaneously articulate and silent, she is fractured by her access to language. Yet we might put things more positively than this. If this is the utterance of a New Woman, its feminism deploys a linguistic strategy which is both very old and very new. On one level, she is an alternative image of Nora in Ibsen's *The Doll's House*, slamming the door against patriarchy and *affirming* an alternative world by saying 'No' to what she has. Her nay-saying is thus an effect of her role as New Woman, and must be read as a positive act. This position has been recoined more recently in the positive refusals of Luce Irigaray's *'Ce sexe qui n'en est pas un'* (this sex which is not *one*, this sex which is not *a sex*), or Julia Kristeva's *'La femme, ce n'est jamais ça'* (woman can never be defined), the title of an interview from 1974 in which Kristeva says:

> a woman cannot "be"; it is something which does not even belong in the order of being. It follows that a feminist practice can only be negative, at odds with what already exists so that we may say "that's not it" and "that's still not it." In "woman" I see something that cannot be represented, something that is not said, something above and beyond nomenclatures and ideologies.[28]

The positive value of negativity is also something about which psychoanalysis has a lot to say. In Chapter 4 I discussed the *a*temporality of the Freudian unconscious – the subject only experiences itself in time consciously. If history and time do not exist in the unconscious, then neither does negativity, which is introduced as a function of repression. Freud discusses this in two extremely important footnotes, hidden in two of the case histories, where he justifies reading the patient's denial in reverse, as a statement of affirmation. In the *Wolf Man* case he writes, 'In the unconscious . . . "No" does not exist, and there is no distinction between contraries. Negation is only introduced by the process of repres-

26 p. 127.
27 p. 167.
28 See Irigaray, 'This Sex Which Is Not One', pp. 99-106; and Kristeva, 'Woman Can Never Be Defined', pp. 137-141, *New French Feminisms*, ed. Marks and de Coutivron, quotation p. 137.

sion'.[29] Similarly, in *Dora* he writes, 'there is no such thing as an unconscious "No"', and argues that in analysis conscious denials need to be read directly as unconscious affirmations:

> an exclamation on the part of the patient of 'I didn't think that', or 'I didn't think of that' . . . can be translated point-blank into: 'Yes, I was unconscious of that.'[30]

Freud builds upon this in the 1925 essay 'Negation'. Here he argues that in analysis certain ideas can only emerge into consciousness in a negative form – that a negative statement is the analysand's way of admitting to something, allowing something through. 'Thus', he writes,

> the content of a repressed image or idea can make its way into consciousness, on condition that it is negated. Negation is a way of taking cognizance of what is repressed; indeed, it is already a lifting of the repression, though not, of course, an acceptance of what is repressed.
>
> . . . To negate something in a judgement is, at bottom, to say: 'This is something which I should prefer to repress.' A negative judgement is the intellectual substitute for repression; its 'no' is the hall-mark of repression, a certificate of origin – like, let us say, 'Made in Germany'.[31]

Were we engaged in a character analysis of Sue here, this would allow us to read in reverse her negatives (*negating* the negative), so that each 'No' about herself becomes a 'Yes'. It seems that in psychoanalysis two negatives *do* make a positive. Hardy had tackled the question of the veracity of sure utterance in another, more politically loaded context in the earlier *Tess of the D'Urbervilles*. Here, the raped heroine addresses her rapist, articulating a version of that feminist rule of thumb which argues that, in matters of conscious control and legal consent, 'Yes means Yes, and No means No': 'Had it never struck your mind', says Tess, 'that what all women say some women might mean?' Here, of course, Hardy is outrageously keeping his options open, the implication being that some women may sometimes *not* say what they mean when it comes to rape. For Freud, and for Hardy too by the time he writes *Jude the Obscure*, 'Yes' doesn't necessarily mean 'Yes', and 'No' certainly doesn't always mean 'No'.

What are the politics of this? A similar issue came into focus in 1992 in Britain with the so-called 'Operation Spanner' case, in which a group of men involved in practising sadomasochism were successfully prosecuted for 'assaulting' each other during sex, even though the parties involved were fully consenting. The law thus overruled the libertarian defence

29 *Wolf Man*, p. 319 n.
30 p. 92 and 92 n.
31 PF vol. xi, p. 437–8.

lobby's argument that in matters of private, consensual sex these things should be possible. Perhaps psychoanalysis muddies the waters even more, but the issues it raises need also to be brought into such a debate. What it highlights about our ability (or inability) to say 'Yes' or 'No' in certain situations only reinforces the need for more clearly demarcated – or perhaps more flexible – boundaries. For whilst psychoanalysis challenges our power of purposeful consent (through its decentring emphasis on the unconscious), by showing how desire (*all* desire) betrays moral codes and confuses how we read (and speak) the simplest of utterances, this is clearly not the evidence the Spanner prosecution brought to court. These are difficult notions, which fundamentally disturb how we think through our own agency. That legal problems of consent only emphasize a question already posed in the humanities stresses the need for a more extensive interdisciplinary debate. This work on the veracity of utterance in literature and psychoanalysis has a central significance for how we develop wider cultural conceptions of the speaking (and consenting) subject.

How does this in turn affect our reading practices? Through this strategy of reversal – as perverse, perhaps, as Sue herself – we are able to move towards a curious position. It seems that it is precisely *because* a text resists that it is most appropriate to a psychoanalytic reading, most appropriable by psychoanalysis. Indeed, from this point of view texts which seem to have most in common thematically with the concerns of psychoanalysis (texts which are manifestly *about*, say, the Oedipus complex or repetition compulsion) may not be as psychoanalytically interesting as texts which resist. Reading Sue Bridehead as Sue Bridehead reads the world (and herself) takes us straight to the heart of one of psychoanalysis's central concerns. Recalcitrance is, then, critically enabling.

Just saying 'No' is only one example of situations in which analysis is able to read against the grain, but there are others. Indeed, in its basic technique psychoanalysis is about overcoming resistance: the point in analysis at which the patient begins to block the process is the point at which the process becomes effective. Resistance (which itself takes a number of forms[32]) is the active defence mechanism which comes into play at exactly the point at which psychoanalysis approaches the truth of what is repressed. The analytic situation thus replays through transference the original moment of repression, the resistances encountered in analysis being 'the same resistances as those which, earlier, made the material concerned into something repressed by rejecting it from the conscious'.[33] Thus repression and resistance enter into an unhappy partnership which analysis must begin to break. In 'The Ego and the Id' (1923), Freud writes:

32 In 'Inhibitions, Symptoms and Anxiety' Freud identifies five types of resistance; see *SE* vol. xx, pp. 318–19.

33 'The Unconscious', PF vol. xi, p. 167.

> The state in which the ideas existed before being made conscious is called by us *repression*, and we assert that the force which instituted the repression and maintains it is perceived as resistance during the work of analysis.[34]

This is one of the cornerstones of psychoanalysis, which makes it unlike any other therapeutic or metapsychological model. It is precisely when its discoveries are being most bitterly challenged that they are being most convincingly proven. Say No to the analysis and you're really saying Yes:

> Only when the resistance is at its height can the analyst, working in common with his patient, discover the repressed instinctual impulses which are feeding the resistance; and it is this kind of experience which convinces the patient of the existence and power of such impulses. . . . If [the doctor] holds fast to this conviction he will often be spared the illusion of having failed when in fact he is conducting the treatment on the right lines.[35]

Something of a leap of faith is required at this stage: it is often when things seem to be going most *wrong* that they're going most *right*.

Freud extends this into a whole theory concerning how and why psychoanalysis was so bitterly opposed during the years of his work. In 'The Resistances to Psychoanalysis' (1925 [1924]) he pictures the public image of his work as concerning an unwelcome psychic and sexual liberation. The assumption that psychoanalysis wants to loosen the fetters of instincts, overturning the primacy of repressed control and allowing the 'ruler' – the ego, whose 'throne rests upon fettered slaves' – to be trampled underfoot, was, he argued, based in fear and a lack of logic. He reads this fear as itself symptomatic: the same model of forces which effect repression in the self are replicated in social responses to psychoanalysis. If analysis tries to 'lift the veil of amnesia from [the] years of childhood', its adult opponents would do their best to lower the veil again in different ways: 'There was only one way out: what psychoanalysis asserted must be false and what posed as a new science must be a tissue of fancies and distortions.'[36]

Here, then, the kind of move Freud makes in his work on civilization and society in the 1920s and 1930s which identifies patterns of individual psychology in models of ancient prehistory (a phylogenetic argument most clearly made in *Totem and Taboo*) is repeated, so that individual psychological models are used to account for bigger social phenomena and historical moves. What happens on a level of individual repression was also, for Freud, occurring on the level of the wider opposition to psychoanalysis itself:

34 PF vol. xi, p. 353.
35 'Remembering, Repeating, Working-Through' (1914), *SE* vol. xii, p. 155.
36 'The Resistances to Psychoanalysis' (1925), PF vol. xv, p. 272.

> The situation obeyed a simple formula: men in the mass behaved to psychoanalysis in precisely the same way as individual neurotics under treatment for their disorders.[37]

Freud's models have clearly had such a currency during the later years of this century that it would be hard to argue that these same resistances prevail. However, since his legacy – particularly in American therapy – was for a long time carried through in the dominant form of 'ego psychology', which was arguably concerned with 'fettering' the 'slaves' of instinct even more, it has been necessary at certain times for writers to highlight once again the important *resistibility* of Freud's original work. For what 'The Resistances to Psychoanalysis' is arguing is that there is, at the heart of the Freudian model, something profoundly unpalatable, uncomfortable, *decentring* of the conscious, moral self. This moral self *must* resist, but its resistance is already a symptom of its fragility, if not its failure. Lacan's so-called 'return to Freud' is essentially about this – by 'saving' analysis from becoming a process which works purely in service of the 'ruling ego', his project was to emphasize the gap, the discontinuities which unbalance the subject. Resistance is necessary.

III
The Uses of Resistance

It could be said, therefore, that psychoanalysis (as opposed to ego psychology) is engaged in a kind of mental guerrilla warfare. It is adept at using to its advantage exactly the forces which would seek to oppose it, deploying – in a sense – the enemy's weapons against it. This subversive strategy has been an important element in the marriage between psychoanalysis and deconstructive reading practices which has been fostered by some writers since the 1970s. A good example of this, as we have seen, is Shoshana Felman's technique of taking from the text the critical terms which are then used to read that text, not as confirmation of its terms but precisely as a way of turning those terms inside out. Thus James's text provides the terms of 'tact' which are then used to read it, as an alternative to confronting it with a vulgar reading which '*tells*'. Future critical practices might take this further. In *Of Grammatology* Jacques Derrida writes:

> The movements of deconstruction do not destroy structures from the outside. They are not possible and effective, nor can they take most accurate aim, except by inhabiting those structures. Inhabiting them *in a certain way*, because

37 Ibid., p. 272.

> one always inhabits, and all the more when one does not suspect it. Operating necessarily from the inside, borrowing all the strategic and economic resources of subversion from the old structure . . .[38]

But this is to put the situation from the perspective of deconstruction, which would 'inhabit in a certain way'. As Derrida tells us at the start of 'Freud and the Scene of Writing', deconstruction is not 'a psychoanalysis of philosophy', and neither is psychoanalysis entirely engaged in a deconstruction of the psyche. Instead of this infiltrating inhabitation, psychoanalysis takes the processes of psychical opposition and allows them to undo themselves – pushing those moments of denial and resistance until the structure which they defend capitulates. The quotation above from *Of Grammatology* seems to me to be most appropriate here, because it suggests that the subversion of boundaries, the politically oriented overthrowing of oppressive jurisdictions, must come from some kind of appropriation of the power of the structure one would overthrow. In that psychoanalysis is already engaged in something of this, it effects similar forms of textual change. Indeed, a remarkably similar sentiment comes, from Che Guevara's seminal statement on infiltration and appropriation, *Guerrilla Warfare*, in which he maintains that the weapons used to overthrow a system must come from within that system: '[T]he principal source of provision for the guerrilla force is precisely the enemy armaments.'[39]

This could stand as a model for how psychoanalytic reading might be set to proceed. I am not suggesting that we posit the text as 'enemy' and take up (its own) arms against it, although this is, in a sense, already implied by the practice which brings theory *to* text and encourages the former to dominate the latter, with text acting then only as fodder for externally constructed critical 'strategies' (a term which implies enmity between text and theory if ever there was one). If Freud is one of Paul Ricœur's 'three masters of suspicion' (along with Nietzsche and Marx), we should be encouraged to become ever more suspicious readers, reading the text 'uncomfortably' in terms of its own movements of 'the hidden and the revealed'. Rather than 'attacking' literature as the material to be conquered, psychoanalytic criticism might nevertheless proceed by taking the resources, terms and unconscious traces of the text, and then turning these back upon it in an act which discloses that which the text would conceal from itself.

38 *Of Grammatology* (Baltimore, MD: Johns Hopkins University Press, 1976), p. 24; his italics.

39 *Guerrilla Warfare*, (Harmondsworth: Penguin, 1969), p. 20.

Bibliography

ADAMS, PARVEEN: 'What Is a Woman? Some Psychoanalytic Dimensions', in *Women: A Cultural Review*, **1/1** (Apr. 1990).

ALTHUSSER, LOUIS: 'Ideology and Ideological State Apparatuses (Notes Towards an Investigation)', in *Lenin and Philosophy and Other Essays*, trans. Ben Brewster. New York: Monthly Review Press, 1971.

APPIGNANESI, LISA and JOHN FORRESTER: *Freud's Women*. London: Weidenfeld & Nicolson, 1992.

AUERBACH, NINA: 'Magi and Maidens: The Romance of the Victorian Freud', in *Writing and Sexual Difference*, ed. Elizabeth Abel, Brighton: Harvester, 1982.

BALDICK, CHRIS: *The Social Mission of English Criticism*. Oxford: Clarendon, 1983.

BALMARY, MARIE: *Psychoanalyzing Psychoanalysis: Freud and the Hidden Fault of the Father*, trans. Ned Lukacher. Baltimore, MD, and London: Johns Hopkins University Press, 1982.

BARNES, JONATHAN: *Early Greek Philosophy*. Harmondsworth: Penguin, 1987.

BARTHES, ROLAND: *Image/Music/Text*, trans. Stephen Heath. London: Fontana, 1977.

BAUDRY, JEAN-LOUIS: 'The Apparatus: Metaphysical Approaches to the Impression of Reality in Cinema', in *Film Theory and Criticism: Introductory Readings*, ed. Gerald Mast, Marshall Cohen and Leo Braudy. Oxford: Oxford University Press, 1992.

BENJAMIN, JESSICA: *The Bonds of Love: Psychoanalysis, Feminism, and the Problem of Domination*. London: Virago, 1990.

BENVENUTO, BICE and ROGER KENNEDY: *The Works of Jacques Lacan*. London: Free Association Books, 1986.

BERGSTROM, JANET, and MARY ANN DOANE (eds.): *The Spectatrix*, special edition of *Camera Obscura*, **20-1** (1989).

BERNHEIMER, CHARLES and CLAIRE KAHANE (eds.): *In Dora's Case: Freud – Hysteria – Feminism*. New York: Columbia University Press, 1985.

BLOOM, HAROLD (ed.): *Edgar Allan Poe*. New York: Chelsea House, 1985.

BONAPARTE, MARIE: *The Life and Works of Edgar Allan Poe: A Psycho-Analytic Interpretation*, trans. John Rodker. London: Imago, 1949.

BOOTHBY, RICHARD: *Death and Desire: Psychoanalytic Theory in Lacan's Return to Freud*. London and New York: Routledge, 1991.

BRENNAN, TERESA (ed.): *Between Feminism and Psychoanalysis*. London and New York, Routledge, 1989.
BRONFEN, ELISABETH: *Over Her Dead Body: Death, Femininity and the Aesthetic*. Manchester: Manchester University Press, 1992.
BRONTË, EMILY: *Wuthering Heights* (1847), ed. Linda H. Peterson. Boston: St Martin's Press, 1992.
BROOKE-ROSE, CHRISTINE: 'Ill Wit and Sick Tragedy: Jude the Obscure', in *Alternative Hardy*, ed. Lance Butler. London and Basingstoke: Macmillan, 1989.
BROOKS, PETER: 'Freud's Masterplot', in *Literature and Psychoanalysis: The Question of Reading: Otherwise*, ed. Shoshana Felman. Baltimore, MD, and London: Johns Hopkins University Press, 1980.
—— 'The idea of a psychoanalytic literary criticism', in *Discourse in Psychoanalysis and Literature*, ed. Shlomith Rimmon-Kenan.
—— *Reading for the Plot: Design and Intention in Narrative*. Oxford: Clarendon, 1984.
BROWNE, NICK: 'The Spectator-in-the-Text: The Rhetoric of *Stagecoach*', in *Movies and Methods ii*, ed. Bill Nichols. Berkeley, CA, and London: University of California Press, 1985.
BUCK, CLAIRE: '"O Careless, Unspeakable Mother": Irigaray, H.D. and Maternal Origin', in *Feminist Criticism: Theory and Practice*, ed. Susan Sellers. Hemel Hempstead: Harvester-Wheatsheaf, 1991.
BURGIN, VICTOR, JAMES DONALD and CORA KAPLAN (eds.): *Formations of Fantasy*. London: Methuen, 1986.
BYARS, JACKIE: 'Gazes/Voices/Power: Expanding Psychoanalysis for Feminist Film and Television Theory', in *Female Spectators: Looking at Film and Television*, ed. Deidre Pribram. London and New York: Verso, 1988.
CAPLAN, PAULA: *The Myth of Women's Masochism*. London: Methuen, 1986.
CARTER, ANGELA: *The Bloody Chamber and Other Stories* (1979). Harmondsworth: Penguin, 1985.
—— *Fireworks* (1974). London: Virago, 1987.
—— *Heroes and Villains* (1969). Harmondsworth: Penguin, 1982.
—— *The Infernal Desire Machines of Doctor Hoffman* (1972). Harmondsworth: Penguin, 1982.
—— *The Magic Toyshop* (1967). London: Virago, 1981.
—— *Nights at the Circus* (1984). London: Chatto & Windus, 1984.
—— *The Passion of New Eve* (1977). London: Virago, 1982.
—— *The Sadeian Woman: An Exercise in Cultural History*. London: Virago, 1979.
—— *Wise Children*. London: Chatto & Windus, 1991.
CHAPMAN, ROWENA and JONATHAN RUTHERFORD (eds.), *Male Order: Unwrapping Masculinity*. London: Lawrence & Wishart, 1988.
CHODOROW, NANCY: *The Reproduction of Mothering: Psychoanalysis and the Sociology of Gender*. Berkeley, CA: University of California Press, 1978.
CIXOUS, HÉLÈNE: 'The Laugh of the Medusa', trans. Keith Cohen and Paula Cohen. In *New French Feminisms*, ed. Elaine Marks and Isabelle de Courtivron, Brighton: Harvester, 1981.
—— and CATHERINE CLÉMENT: *The Newly Born Woman*, trans. Betsy Wing. Manchester: Manchester University Press, 1986.
CLÉMENT, CATHERINE: *The Lives and Legends of Jacques Lacan*, trans. Arthur Goldhammer. New York: Columbia, 1983.

Clover, Carol J.: *Men, Women and Chain Saws: Gender in the Modern Horror Film*. London: BFI, 1992.
Cook, Pam and Claire Johnston: 'The Place of Woman in the Cinema of Raoul Walsh', collected in *Feminist Film Theory*, ed. Constance Penly. London and New York: BFI and Routledge, 1988.
Derrida, Jacques: *Positions*, trans. Alan Bass. London: Athlone, 1981.
—— *Of Grammatology*, trans. Gayatri Spivak. Baltimore and London: Johns Hopkins University Press, 1976.
—— 'The Purveyor of Truth', in *The Post Card: From Socrates to Freud and Beyond*, trans. Alan Bass. Chicago: Chicago University Press, 1987; also collected in *The Purloined Poe*, ed. Muller and Richardson.
Doane, Mary Ann: *The Desire to Desire: The Woman's Film of the 1940s*. Bloomington: Indiana University Press, 1987.
Donald, James (ed.): *Fantasy and the Cinema*. London: BFI, 1989.
—— (ed.): *Psychoanalysis and Cultural Theory: Thresholds*. London: Macmillan, 1991.
Dyer, Richard: 'Don't Look Now: The Male Pin-Up', in *The Sexual Subject*, ed. *Screen*. London and New York: Routledge, 1992.
Eagleton, Terry: *Critical Theory: An Introduction*. London and Basingstoke: Macmillan, 1983.
Felman, Shoshana (ed.): *Literature and Psychoanlaysis: The Question of Reading: Otherwise*. Baltimore, MD, and London: Johns Hopkins University Press, 1980.
—— 'On Reading Poetry: Reflections on the Limits and Possibilities of Psychoanalytical Approaches', collected in Felman's 1980 volume, *The Literary Freud: Mechanisms of Defense and the Poetic Will*; also collected in *The Purloined Poe*, ed. Muller and Richardson.
Fenichel, Otto: 'Critique of the Death Instinct' (1935), in *The Collected Papers of Otto Fenichel*, ed. Hanna Fenichel and David Rapaport. London: Routledge & Kegan Paul, 1954.
Ferenczi, Sándor: 'Confusion of Tongues between Adults and the Child ('The Language of Tenderness and the Language of [Sexual] Passion') (1932), trans. Jeffrey M. Masson and Marianne Loring, in Masson, *The Assault on Truth*.
Foucault, Michel: *Madness and Civilization: A History of Insanity in the Age of Reason* (1961), trans. Richard Howard. London: Tavistock, 1967.
Frayling, Christopher: *Vampyres: Lord Byron to Count Dracula*. London and Boston: Faber & Faber, 1991.
Freud, Sigmund: 'The Aetiology of Hysteria' (1896), in *The Assault on Truth: Freud's Suppression of the Seduction Theory* by J. M. Masson, trans. James Strachey. Harmondsworth: Penguin, 1985.
—— 'Analysis of a Phobia in a Five-Year-Old Boy ("Little Hans")' (1909), in *Case Histories I*, trans. Alix and James Strachey, ed. Angela Richards, Pelican Freud Library (hereafter PF) vol. viii. Harmondsworth: Penguin, 1980.
—— 'Analysis Terminable and Interminable' (1937), in *The Standard Edition of the Complete Psychological Works of Sigmund Freud* (London: Hogarth Press, 1955–74; hereafter *SE*), vol. xxiii.
—— 'Beyond the Pleasure Principle' (1920), in *On Metapsychology*, trans. James Strachey, ed. Angela Richards, PF vol. xi. Harmondsworth: Penguin, 1984.
—— 'A Case of Paranoia Running Counter to the Psychoanalytic Theory of the Disease' (1915), in *On Psychopathology: Inhibitions, Symptoms and Anxiety and*

other works, trans. James Strachey, ed. Angela Richards, PF vol. x. Harmondsworth: Penguin, 1981.

FREUD, SIGMUND: '"A Child is Being Beaten"(A Contribution to the Study of the Origin of the Sexual Perversions)' (1919), in PF vol. x. Harmondsworth: Penguin, 1981.

—— 'Civilization and its Discontents' (1930 [1929]), in *Civilization, Society and Religion*, trans. James Strachey, ed. Albert Dickson, PF vol. xii. Harmondsworth: Penguin, 1985.

—— *The Complete Letters of Sigmund Freud to Wilhelm Fliess 1887–1904*, trans. and ed. Jeffrey Moussaieff Masson. Cambridge, MA, and London: Harvard University Press, 1985.

—— 'Creative Writers and Day-dreaming' (1908 [1907]), in *Art and Literature*, trans. James Strachey, ed. Albert Dickson, PF vol. xiv. Harmondsworth: Penguin, 1985.

—— 'The Economic Problem of Masochism' (1924), in PF vol. xi. Harmondsworth: Penguin, 1984.

—— 'The Ego and the Id' (1923), in PF vol. xi. Harmondsworth: Penguin, 1984.

—— 'Fetishism' (1927), in *On Sexuality*, trans. James Strachey, ed. Angela Richards, PF vol.vii, Harmondsworth: Penguin, 1977.

—— 'Fragment of an Analysis of a Case of Hysteria ('Dora') (1905 [1901]), in *Case Histories I*, trans. Alix and James Strachey, ed. Angela Richards, PF vol. viii.

—— 'From the History of an Infantile Neurosis (the "Wolf Man")', in *Case Histories II*, trans. James Strachey, ed. Angela Richards, PF vol. ix, Harmondsworth: Penguin, 1981.

—— 'Group Psychology and the Analysis of the Ego' (1921), in PF vol. xii. Harmondsworth: Penguin, 1985.

—— 'Instincts and Their Vicissitudes' (1915), in PF vol. xi. Harmondsworth: Penguin, 1984.

—— *The Interpretation of Dreams* (1900), trans. James Strachey, ed. Angela Richards, PF vol. iv. Harmondsworth: Penguin, 1982.

—— *Introductory Lecture on Psychoanalysis* (1916-17 [1915-17]), trans. James Strachey, ed. Strachey and Angela Richards, PF vol. i. Harmondsworth: Penguin, 1979.

—— 'Leonardo da Vinci and a Memory of His Childhood' (1910), in PF vol. xiv. Harmondsworth: Penguin, 1985.

—— 'Negation' (1925), in PF vol. xi. Harmondsworth: Penguin, 1984.

—— *New Introductory Lectures on Psychoanalysis* (1933 [1932]), trans. James Strachey, ed. Strachey and Angela Richards, PF vol. ii. Harmondsworth: Penguin, 1981.

—— 'A Note Upon the "Mystic Writing Pad"' (1925 [1924]), in PF vol. xi. Harmondsworth: Penguin, 1984.

—— 'Notes upon a Case of Obsessional Neurosis (the "Rat Man")' (1909), in PF vol. ix. Harmondsworth: Penguin, 1981.

—— 'The Occurrence in Dreams of Material from Fairy Tales' (1912), *SE* vol. xii.

—— 'On Narcissism: An Introduction' (1914), in PF vol. xi. Harmondsworth: Penguin, 1984.

—— *The Origins of Psycho-Analysis: Letters to Wilhelm Fliess, Drafts and Notes, 1887–1902*, ed. Marie Bonaparte, Anna Freud and Ernst Kris, trans. Eric Mosbacher and James Strachey. New York: Basic Books, 1977.

—— 'An Outline of Psychoanalysis' (1940 [1938]), in *Historical and Expository Works on Psychoanalysis*, trans. James Strachey, ed. Albert Dickson, PF vol. xv. Harmondsworth: Penguin, 1986.
—— 'Psychoanalysis', in 'Two Encyclopaedia Articles' (1923 [1922]), in PF vol. xv. Harmondsworth: Penguin, 1986.
—— 'Psychoanalytic Notes on An Autobiographical Account of A Case of Paranoia (Dementia Paranoides) (Schreber)', in PF vol. ix. Hamondsworth: Penguin, 1981.
—— 'Psychopathic Characters on the Stage' (1942 [1905–6]), in PF vol. xiv. Harmondsworth: Penguin, 1985.
—— 'The Question of Lay Analysis' (1926), in PF vol. xv. Harmondsworth: Penguin, 1986.
—— 'Remembering, Repeating, Working-Through', in *SE* vol. xii.
—— 'The Resistances to Psychoanalysis' (1925 [1924]), in PF vol. xv. Harmondsworth: Penguin, 1986.
—— 'A Short Account of Psychoanalysis' (1924 [1923]) in PF vol. xv. Harmondsworth: Penguin, 1986.
—— 'The Splitting of the Ego in the Process of Defence' (1940 [1938]), in PF vol. xi. Harmondsworth: Penguin, 1984.
—— 'Thoughts for the Times on War and Death' (1915), in PF vol. xii. Harmondsworth: Penguin, 1985.
—— 'Three Essays on the Theory of Sexuality' (1905), in PF vol. vii. Harmondsworth: Penguin, 1977.
—— 'Totem and Taboo' (1913 [1912-13]), in *The Origins of Religion*, trans. James Strachey, ed. Albert Dickson, PF vol. xiii, Harmondsworth: Penguin, 1985.
—— 'The "Uncanny"' (1919), in *Art and Literature*, PF vol. xiv. Harmondsworth: Penguin, 1985.
—— 'The Unconscious' (1915), in PF vol. xi. Harmondsworth: Penguin, 1984.
—— and JOSEPH BREUER: *Studies on Hysteria*, trans. James and Alix Strachey, ed. Angela Richards, PF vol. iii. Harmondsworth: Penguin, 1980.
GALLOP, JANE: *Feminism and Psychoanalysis: The Daughter's Seduction*. London and Basingstoke: Macmillan, 1982.
—— 'Keys to Dora', in *In Dora's Case*, ed. Bernheimer and Kahane.
—— *Reading Lacan*. Ithaca, NY, and London: Cornell University Press, 1985.
GAMMAN, LORRAINE and MARGARET MARSHMENT (eds.): *The Female Gaze: Women as Viewers of Popular Culture*. London: Women's Press, 1988.
GARDINER, JUDITH KEGAN: 'Mind Mother: Psychoanalysis and Feminism', in *Making a Difference*, ed. Greene and Kahn. London and New York: Routledge, 1985.
GARDINER, MURIEL (ed.): *The Wolf-Man and Sigmund Freud*. London: Karnac, 1989.
GARNER, SHIRLEY NELSON, CLAIRE KAHANE and MADELON SPRENGNETHER (eds.): *The (M)Other Tongue: Essays in Feminist Psychoanalytic Interpretation*. Ithaca, NY, and London: Cornell University Press, 1986.
GASKELL, ELIZABETH: *Cousin Phillis* (1864). Harmondsworth: Penguin, 1986.
GILBERT, SANDRA and SUSAN GUBAR: *The Madwoman in the Attic*. New Haven, CT, and London: Yale University Press, 1979.
GILMAN, CHARLOTTE PERKINS: *The Yellow Wallpaper* (1899). London: Virago, 1981.
GIRARD, RENÉ: *Deceit, Desire and the Novel: Self and Other in Literary Structure* (1961), trans. Yvonne Freccero. Baltimore, MD, and London: Johns Hopkins University Press, 1969.

GOODE, JOHN: 'Sue Bridehead and the New Woman', in *Women Writing and Writing About Women*, ed. Jacobus.
GREENE, GAYLE and COPPÉLIA KAHN (eds.): *Making a Difference: Feminist Literary Criticism*. London: Methuen, 1985.
GUNN, DANIEL: *Psychoanalysis and Fiction: An Exploration of Literary and Psychoanalytic Borders*. Cambridge: Cambridge University Press, 1990.
HANEY-PERITZ, JANICE: 'Monumental Feminism and Literature's Ancestral House: Another Look at *The Yellow Wallpaper*', *Women's Studies*, **12** (1986).
HARDY, THOMAS: *Jude the Obscure* (1896). London and Basingstoke: Macmillan, 1978.
—— *Tess of the D'Urbervilles* (1891). Harmondsworth: Penguin, 1990.
HARTMAN, GEOFFREY (ed.): *Psychoanalysis and the Question of the Text: Selected Papers from the English Institute, 1976-77*. Baltimore, MD: Johns Hopkins University Press, 1978.
HEIDEGGER, MARTIN: *Nietzsche*, ii: *The Eternal Recurrence of the Same*, trans. David Farrell Krell. San Francisco, CA, 1984.
HERTZ, NEIL: 'Dora's Secrets, Freud's Techniques', in *In Dora's Case*, ed. Bernheimer and Kahane.
—— 'Freud and the Sandman', in *Textual Strategies*, ed. Josué V. Harari. Ithaca, NY: Cornell University Press, 1979.
HOLLAND, NORMAN: *Five Readers Reading*. New Haven, CT: Yale University Press, 1975.
HUNTER, DIANNE: 'Hysteria, Psychoanalysis, and Feminism: The Case of Anna O', *Feminist Studies*, **9/3**. (Fall, 1983).
IRIGARAY, LUCE: 'And the One Doesn't Stir Without the Other', trans. Hélène Vivienne Wenzel, *Signs* (Autumn 1981).
—— 'This Sex Which Is Not One', trans. Claudia Reeder, in *New French Feminisms*, ed. Marks and de Courtivron.
JACOBUS, MARY: 'Sue the Obscure', *Essays in Criticism*, **25** (July 1975).
—— (ed.): *Women Writing and Writing About Women*. London: Croom Helm, 1979.
JACOBY, RUSSELL: *The Repression of Psychoanalysis: Otto Fenichel and the Political Freudians*. New York: Basic Books, 1983.
JACKSON, ROSEMARY: *Fantasy: The Literature of Subversion*. London and New York: Methuen, 1981.
JAMES, HENRY: *The Turn of the Screw and other stories*. Harmondsworth: Penguin, 1982.
JAWORZYN, STEFAN (ed.): *Shock Xpress*. London: Titan, 1991.
JOHNSON, BARBARA: *The Critical Difference: Essays in the Contemporary Rhetoric of Reading*. Baltimore, MD, and London: Johns Hopkins University Press, 1980.
JONES, ANN ROSALIND: 'Inscribing Femininity: French Theories of the Feminine', in *Making a Difference*, ed. Greene and Kahn.
JONES, ERNEST: *The Life and Work of Sigmund Freud* (1961). Harmondsworth: Penguin, 1984.
—— *On the Nightmare*. London: Hogarth Press, 1931.
JORDAN, ELAINE: 'The Dangers of Angela Carter', in *New Feminist Discourses*, ed. Isobel Armstrong. London and New York: Routledge, 1992.
KAPLAN, E. ANN (ed.): *Psychoanalysis and Cinema*. New York and London: Routledge, 1990.
KASMER, LISA: 'Charlotte Perkins Gilman's "The Yellow Wallpaper": A Symptomatic Reading', *Literature and Psychology*, **36/3** (1990).

KAWIN, BRUCE: 'The Mummy's Pool', in *Film Theory and Criticism: Introductory Readings*, ed. Gerald Mast, Marshall Cohen and Leo Braudy. Oxford: Oxford University Press, 1992.
KENNARD, JEAN E.: 'Convention Coverage or How to Read Your Own Life', *New Literary History*, **13/1** (Autumn 1981).
KING, STEPHEN: *Danse Macabre*. London and Basingstoke: Macmillan, 1992.
KLEIN, MELANIE: *The Psycho-Analysis of Children* (1932), trans. Alix Strachey. London: Hogarth Press, 1980.
—— *The Selected Melanie Klein*, ed. Juliet Mitchell. Harmondsworth: Penguin, 1986.
KOLODNY, ANNETTE: 'A Map for Rereading: Or Gender and the Interpretation of Literary Texts', *New Literary History*, **11/3** (Spring 1980).
KRIS, ERNST: *Psychoanalytic Exploration in Art*. New York: International Universities Press, 1952.
KRISTEVA, JULIA: *Desire in Language: A Semiotic Approach to Literature and Art*, trans. Thomas Gora, Alice Jardine and Leon S. Roudiez. Oxford: Blackwell, 1984.
—— 'Woman Can Never Be Defined', trans. Marilyn A. August, in *New French Feminisms*, ed. Marks and de Courtivron.
KRUTCH, JOSEPH WOOD: *Edgar Allan Poe: A Study in Genius*. New York: Knopf, 1926.
KUTTNER, ALFRED BOOTH: '*Sons and Lovers*: A Freudian Appreciation' (1916), in *'Sons and Lovers': A Casebook*, ed Gāmini Salgādo. London: Macmillan, 1975.
LACAN, JACQUES: *Écrits: A Selection* (1966), trans. Alan Sheridan. London: Tavistock, 1980.
—— *The Four Fundamental Concepts of Psycho-Analysis* (1973), trans. Alan Sheridan. New York and London: Norton, 1981.
—— 'Intervention on Transference', in *In Dora's Case*, ed. Bernheimer and Kahane.
—— *Speech and Language in Psychoanalysis* (1968), trans. Anthony Wilden. Baltimore, MD, and London: Johns Hopkins University Press, 1981.
—— and the *École Freudienne: Feminine Sexuality*, ed. Juliet Mitchell and Jacqueline Rose, trans. Jacqueline Rose. London and Basingstoke: Macmillan, 1982.
LAPLANCHE, JEAN: *Life and Death in Psychoanalysis* (1970), trans. Jeffrey Mehlman. Baltimore, MD, and London: Johns Hopkins University Press, 1990.
—— *New Foundations for Psychoanalysis* (1987), trans. David Macey. Oxford: Blackwell, 1989.
—— *Seduction, Translation, Drives*, trans. Martin Stanton. London: ICA, 1992.
—— and J.-B. PONTALIS: *The Language of Psycho-Analysis*, trans. Donald Nicholson-Smith. London: Hogarth Press, 1983.
—— 'Fantasy and the Origins of Sexuality', *International Journal of Psycho-Analysis*, **49/1** (1968); also collected in *Formations of Fantasy*, ed. Burgin *et al.*
LAWRENCE, D. H.: *Apocalypse* (1931). Harmondsworth: Penguin, 1979.
—— *Fantasia of the Unconscious/Psychoanalysis and the Unconscious* (1923/1921). Harmondsworth: Penguin, 1977.
—— *A Selection from Phoenix*, ed. A. A. H. Inglis. Harmondsworth: Penguin, 1979.
—— *Sons and Lovers* (1913). Harmondsworth: Penguin, 1978.
LEIGHTON, ANGELA: '"When I am dead, my dearest": The Secret of Christina Rossetti', *Modern Philology*, **87** (1990).

LE FANU, J. SHERIDAN: 'Carmilla' (1872), in *The Penguin Book of Vampire Stories*, ed. Alan Ryan. London: Bloomsbury, 1991.

LESSER, SIMON O.: *Fiction and the Unconscious*. Chicago: Chicago University Press, 1957.

MACCABE, COLIN (ed.): *The Talking Cure: Essays in Psychoanalysis and Language*. London and Basingstoke: Macmillan, 1981.

MACCANNELL, JULIET FLOWER: *Figuring Lacan: Criticism and the Cultural Unconscious*. London and Sydney: Croom Helm, 1986.

MAHONY, PATRICK: *Freud as a Writer*. New York: International Universities Press, 1982.

MALCOLM, JANET: *In the Freud Archives*. London: Fontana, 1986.

MARCUS, MARIA: *A Taste For Pain: On Masochism and Female Sexuality*, trans. Joan Tate. London: Souvenir Press, 1981.

MARCUS, STEVEN: 'Freud and Dora: Story, History, Case History', in *In Dora's Case*, ed. Bernheimer and Kahane.

MARKS, ELAINE and ISABELLE DE COURTIVRON (eds.): *New French Feminisms*. Brighton: Harvester, 1981.

MASSON, J. M.: *The Assault on Truth: Freud's Suppression of the Seduction Theory*. Harmondsworth: Penguin, 1985.

MAYNE, JUDITH: *Cinema and Spectatorship*. London and New York: Routledge, 1993.

MCGANN, JEROME J.: 'The Religious Poetry of Christina Rossetti', in *The Beauty of Inflections: Literary Investigations in Historical Method and Theory*. Oxford: Clarendon Press, 1985.

MEISEL, PERRY (ed.) : *Freud: A Collection of Critical Essays*. Englewood Cliffs, NJ: Prentice-Hall, 1981.

MELLARD, JAMES: *Using Lacan, Reading Fiction*. Urbana, IL: University of Illinois Press, 1991.

METZ, CHRISTIAN: *Psychoanalysis and Cinema: The Imaginary Signifier*, trans. Celia Britton, Annwyl Williams, Ben Brewster and Alfred Guzzetti. London and Basingstoke: Macmillan, 1990.

MIDDLETON, PETER: *The Inward Gaze: Masculinity and Subjectivity in Modern Culture*. London and New York: Routledge, 1992.

MILLETT, KATE: 'Beyond Politics? Children and Sexuality', in *Pleasure and Danger: Exploring Female Sexuality*, ed. Vance.

MITCHELL, JULIET: *Psychoanalysis and Feminism* (1974). Harmondsworth: Penguin, 1982.

— *Women – The Longest Revolution: Essays on Feminism, Literature and Psychoanalysis*. London: Virago, 1984.

MODLESKI, TANIA: 'Time and Desire in the Woman's Film', in *Film Theory and Criticism: Introductory Readings*, ed. Gerald Mast, Marshall Cohen and Leo Braudy. Oxford: Oxford University Press, 1992.

MORGAN, ROSEMARIE: *Women and Sexuality in the Novels of Thomas Hardy*. London and New York: Routledge, 1988.

MORRILL, DAVID F.: '"Twilight is not good for maidens": Uncle Polidori and the Psychodynamics of Vampirism in "Goblin Market"', *Victorian Poetry*, **28/1**.

MORRISON, TONI: *Beloved*. London: Picador, 1988.

MULHERN, FRANCIS: *The Moment of Scrutiny*. London: New Left Books, 1979.

MULLER, JOHN P. and WILLIAM J. RICHARDSON (eds.): *The Purloined Poe: Lacan,*

Derrida, and Psychoanalytic Reading, Baltimore and London: Johns Hopkins University Press, 1988.

MULVEY, LAURA: 'Visual Pleasure and Narrative Cinema', and 'Afterthoughts on "Visual Pleasure and Narrative Cinema" inspired by *Duel in the Sun*', in *Visual and Other Pleasures*. Bloomington: Indiana University Press, 1989).

NEALE, STEVEN: 'Masculinity as Spectacle', in *The Sexual Subject*, ed. *Screen*. London and New York: Routledge, 1992.

NEWMAN, KIM: *Nightmare Movies*. London: Bloomsbury, 1988.

NIETZSCHE, FRIEDRICH: *Also Sprach Zarathustra*, in *Werke*, vol. ii, ed. Karl Schlechta. Munich: 1981.

OBHOLZER, KARIN: *The Wolf-Man, Sixty Years Later: Conversations with Freud's Controversial Patient*, trans. Michael Shaw. London: Routledge & Kegan Paul, 1982.

OLIPHANT, MRS: *The Autobiography of Mrs. Oliphant* (1899). Chicago and London: University of Chicago Press, 1988.

PENLEY, CONSTANCE (ed.): *Feminism and Film Theory*. London and New York: Routledge/BFI, 1988.

POE, EDGAR ALLAN: *Selected Writings*. Harmondsworth: Penguin, 1984.

POOLE, ROGER: 'Psychoanalytic Theory: D. H. Lawrence: *St Mawr*', in *Literary Theory at Work: Three Texts*, ed. Douglas Tallack. London: Batsford, 1987.

PRIBRAM, E. DEIDRE (ed.): *Female Spectators: Looking at Film and Television*. London and New York: Verso, 1988.

REICH, WILHELM: *Reich Speaks of Freud*. Harmondsworth: Penguin, 1975.

RICH, ADRIENNE: *Of Woman Born: Motherhood as Experience and Institution*. London: Virago, 1977.

RIMMON-KENAN, SHLOMITH (ed.): *Discourse in Psychoanalysis and Literature*. London and New York: Methuen, 1987.

RIVIERE, JOAN: 'Womanliness as a Masquerade' (1929), in *Formations of Fantasy*, ed. Burgin *et al.*, London and New York: Methuen, 1986.

ROSE, JACQUELINE: *Sexuality in the Field of Vision*. London: Verso, 1986.

—— *Why War? Psychoanalysis, Politics, and the Return to Melanie Klein*. Oxford: Blackwell, 1993.

ROSSETTI, CHRISTINA: *Selected Poems*, ed. C. H. Sisson. Manchester: Carcanet, 1984.

ROUSSEL, JEAN: 'Introduction to Jacques Lacan', *New Left Review*, **51** (Sept.–Oct. 1968).

ROUSTANG, FRANÇOIS: *Dire Mastery: Discipleship from Freud to Lacan*, trans. Ned Lukacher. Baltimore, MD: Johns Hopkins University Press, 1982.

ROYLE, NICHOLAS and ANN WORDSWORTH (eds.): *Psychoanalysis and Literature: New Work*, special edition of *The Oxford Literary Review*, **12/1–2** (1990).

SAGE, LORNA: 'Death of the Author', in *Granta* **41**, 'Biography'.

—— 'A Savage Sideshow', *New Review* **4/39–40** (July 1977).

Screen (ed.): *The Sexual Subject: A 'Screen' Reader in Sexuality*. London and New York: Routledge, 1992.

SEGAL, HANNA: *Introduction to the Work of Melanie Klein*. London: Karnac, 1988.

SHOWALTER, ELAINE: *The Female Malady: Women, Madness and English Culture, 1830-1980*. London: Virago, 1987.

SILVERMAN, KAJA: *Male Subjectivity at the Margins*. New York and London: Routledge, 1992.

SKURA, MEREDITH ANNE: *The Literary Use of the Psychoanalytic Process*. New Haven, CT, and London: Yale University Press, 1981.

SMITH, JOSEPH H. (ed.): *The Literary Freud: Mechanisms of Defense and the Poetic Will.* New Haven, CT: Yale University Press, 1980.

SMITH, JOSEPH H. and WILLIAM KERRIGAN (eds.): *Interpreting Lacan.* New Haven, CT, and London: Yale University Press, 1983.

SPRENGNETHER, MADELON: 'Enforcing Oedipus: Freud and Dora', in *In Dora's Case,* ed. Bernheimer and Kahane.

STACEY, JACKIE: *Star Gazing: Hollywood Cinema and Female Spectatorship.* London and New York: Routledge, 1994.

STALLYBRASS, PETER and ALLON WHITE: *The Politics and Poetics of Transgression.* London: Methuen, 1986.

STIMPSON, CATHARINE R.: 'Zero Degree Deviancy: The Lesbian Novel in English', in *Writing and Sexual Difference,* ed. Elizabeth Abel. Brighton: Harvester, 1982.

STOKER, BRAM: *Dracula* (1897). Harmondsworth: Penguin, 1979.

TODOROV, TZVETAN: *The Fantastic: A Structural Approach to a Literary Genre,* trans. Richard Howard. Ithaca, NY: Cornell University Press, 1973.

TREMAIN, ROSE: *Sacred Country.* London: Hodder & Stoughton, 1992.

TRILLING, LIONEL: 'Freud and Literature', in *Freud,* ed. Meisel, pp. 95–111.

VANCE, CAROLE S. (ed.): *Pleasure and Danger: Exploring Female Sexuality.* Boston and London: Routledge & Kegan Paul, 1984.

VEITH, ILZA: *Hysteria: The History of a Disease.* Chicago: Chicago University Press, 1965.

WALKER, ALICE: *In Search of Our Mother's Gardens and Other Essays.* London: Women's Press, 1984.

WEISS, ANDREA: *Vampires and Violets: Lesbians in the Cinema.* London: Jonathan Cape, 1992.

WILLIAMS, LINDA: 'When the Woman Looks', in *Film Theory and Criticism: Introductory Readings,* ed. Gerald Mast, Marshall Cohen and Leo Braudy. Oxford: Oxford University Press, 1992.

WILLIAMS, LINDA RUTH: 'Critical Warfare and Henry Miller's *Tropic of Cancer*', in *Feminist Criticism: Theory and Practice,* ed. Susan Sellers. Hemel Hempstead: Harvester-Wheatsheaf, 1991.

—— 'Happy Families? Feminist Reproduction and Matrilineal Thought', in *New Feminist Discourses,* ed. Isobel Armstrong. London and New York: Routledge, 1992.

—— *Sex in the Head: Visions of Femininity and Film in D. H. Lawrence.* Hemel Hempstead: Harvester–Wheatsheaf, 1993.

—— 'Submission and Reading: Feminine Masochism and Feminist Criticism', *New Formations,* special issue, 'Modernism/Masochism' (Spring 1989).

WILSON, EDMUND: *The Triple Thinkers.* Harmondsworth: Penguin, 1962.

WINNICOTT, D. W.: *Playing and Reality* (1971). Harmondsworth: Penguin, 1988.

WOOLF, VIRGINIA: *A Room of One's Own* (1929). St Albans: Granada, 1977.

WRIGHT, ELIZABETH: *Psychoanalytic Criticism: Theory in Practice.* London and New York: Methuen, 1984.

—— (ed.): *Feminism and Psychoanalysis: A Critical Dictionary.* Oxford: Blackwell, 1992.

WRIGHT, T.R.: *Hardy and the Erotic.* London and Basingstoke: Macmillan, 1989.

ŽIŽEK, SLAVOJ: 'The Detective and the Analyst', *Literature and Psychology,* **36/4** (1990).

Index